Molecular Neurobiology

The Practical Approach Series

SERIES EDITORS

D. RICKWOOD
Department of Biology, University of Essex
Wivenhoe Park, Colchester, Essex CO4 3SQ, UK

B. D. HAMES
Department of Biochemistry and Molecular Biology, University of Leeds
Leeds LS2 9JT, UK

Affinity Chromatography
Animal Cell Culture
Animal Virus Pathogenesis
Antibodies I and II
Biochemical Toxicology
Biological Membranes
Biosensors
Carbohydrate Analysis
Cell Growth and Division
Cellular Calcium
Cellular Neurobiology
Centrifugation (2nd edition)
Clinical Immunology
Computers in Microbiology
Cytokines
Directed Mutagenesis
DNA Cloning I, II, and III
Drosophila
Electron Microscopy in Molecular Biology
Electron Microscopy of Tissues and Cells
Essential Molecular Biology I and II
Fermentation
Flow Cytometry
Gel Electrophoresis of Nucleic Acids (2nd edition)
Gel Electrophoresis of Proteins (2nd edition)
Genome Analysis
HPLC of Small Molecules
HPLC of Macromolecules
Human Cytogenetics
Human Genetic Diseases
Immobilised Cells and Enzymes
Iodinated Density Gradient Media
Light Microscopy in Biology
Liposomes
Lymphocytes
Lymphokines and Interferons
Mammalian Cell Biotechnology
Mammalian Development
Medical Bacteriology
Medical Mycology
Microcomputers in Biology
Microcomputers in Physiology
Mitochondria

Molecular Neurobiology
Mutagenicity Testing
Neurochemistry
Nucleic Acid and Protein
  Sequence Analysis
Nucleic Acids Hybridisation
Nucleic Acids Sequencing
Oligonucleotide Synthesis
Oligonucleotides and
  Analogues
Peptide Hormone Action
Peptide Hormone Secretion
Photosynthesis: Energy
  Transduction
Plant Cell Culture
Plant Molecular Biology
Plasmids
Polymerase Chain Reaction
Post-implantation Mammalian
  Development
Prostaglandins and Related
  Substances
Protein Architecture
Protein Function
Protein Purification
  Applications
Protein Purification Methods
Protein Sequencing
Protein Structure
Proteolytic Enzymes
Radioisotopes in Biology
Receptor Biochemistry
Receptor–Effector Coupling
Ribosomes and Protein
  Synthesis
Solid Phase Peptide Synthesis
Spectrophotometry and
  Spectrofluorimetry
Steroid Hormones
Teratocarcinomas and
  Embryonic Stem Cells
Transcription and Translation
Virology
Yeast

Molecular Neurobiology

A Practical Approach

Edited by
JOHN CHAD
and
HOWARD WHEAL

*Department of Neurophysiology,
University of Southampton,
Bassett Crescent East, Southampton, UK*

at
OXFORD UNIVERSITY PRESS
Oxford New York Tokyo

Oxford University Press, Walton Street, Oxford OX2 6DP
Oxford New York Toronto
Delhi Bombay Calcutta Madras Karachi
Petaling Jaya Singapore Hong Kong Tokyo
Nairobi Dar es Salaam Cape Town
Melbourne Auckland
and associated companies in
Berlin Ibadan

Oxford is a trade mark of Oxford University Press

Published in the United States
by Oxford University Press, New York

© Oxford University Press 1991

All rights reserved. No part of this publication may be reproduced,
stored in a retrieval system, or transmitted, in any form or by any means,
electronic, mechanical, photocopying, recording, or otherwise, without
the prior permission of Oxford University Press

This book is sold subject to the condition that it shall not, by way
of trade or otherwise, be lent, re-sold, hired out or otherwise circulated
without the publisher's prior consent in any form of binding or cover
other than that in which it is published and without a similar condition
including this condition being imposed on the subsequent purchaser

British Library Cataloguing in Publication Data
Molecular neurobiology.
1. Neurobiology. Molecular biology
I. Chad, John II. Wheal, Howard III. Series
591.188
ISBN 0–19–963108–5
ISBN 0–19–963109–3 (pbk)

Library of Congress Cataloging in Publication Data
Molecular neurobiology: a practical approach/edited by John Chad
and Howard Wheal.
p. cm.—(Practical approach series)
Companion v. to: Cellular neurobiology.
1. Molecular neurobiology—Methodology. I. Chad, John.
II. Wheal, H. V. III. Series.
[DNLM: 1. Molecular Biology—Methods. 2. Nervous System—
physiology. 3. Neurobiology—methods. WL 25 M7184]
QP356.2.M644 1991
DNLM/DLC 90–15629
for Library of Congress CIP
ISBN 0–19–963108–5 (h/b)
ISBN 0–19–963109–3 (pbk.)

Typeset by Cambrian Typesetters, Frimley, Surrey
Printed in Great Britain by
Information Press Ltd, Eynsham, Oxford

Preface

IN editing the two volumes of *Cellular and molecular neurobiology: a practical approach*, we have attempted to bring together chapters which describe some of the large number of approaches which are being applied in neurobiological research. A comprehensive compendium of all the methodologies currently in use would take tens of volumes; however, in the limited space available we have tried to give a flavour of some of the more interesting and widely applicable approaches which have been recently developed.

We have selected a variety of techniques which will appeal to the new and established researcher. Each chapter is intended to give information which is of direct practical assistance in designing and undertaking new experiments. Practical procedures which can be closely defined are presented as separate protocols within the relevant chapters to facilitate actual use in the laboratory.

The distribution of chapters between the volumes has been based upon the concept that neurobiology consists of the study of processes across a spectrum of size scales, from the whole animal to the single molecule. Thus, in *Cellular neurobiology* we have included approaches to the isolation and maintenance of cells in vitro; electrophysiological approaches to whole cell recordings and isolation of ionic currents; application of optical methods to changing and measuring concentrations of ions or molecules within cells; heuristic modelling of neuronal properties. The latter section relates to attempts to collect together the information from reductionist studies into a synthesis of the properties of intact cells, allowing the comparison of a theoretical understanding with actual properties.

The second volume, *Molecular neurobiology* contains chapters which principally relate to approaches directed at discerning the function of molecules, including sections on recording the properties of ionic channels; second messenger systems; and finally molecular biological approaches. We have searched in vain for an alternative term to clarify the distinction between 'molecular biology' and the 'biology of molecules'. Therefore, we have chosen to use the adjective molecular in its broadest sense to incorporate studies which directly address the behaviour of individual molecules as opposed to the averaged behaviour of systems in cells.

We hope that we have succeeded in our attempt to provide an interesting and useful collection of some of the many approaches now available to the neurobiologist. We would like to thank all the contributors and their colleagues. These chapters represent the product of years of development work, and without their generosity in sharing their experience these volumes could not be published.

Contents

List of contributors   xvii

Abbreviations   xix

IONIC CHANNEL RECORDINGS

1. Progress in instrumentation technology for recording from single channels and small cells   3

Alan S. Finkel

1. Introduction	3
2. Background	3
What is a patch clamp?	3
The importance of a good seal	6
A brief history	6
3. Resistor feedback technology	7
Description of the current-to-voltage head stage	7
Problems	8
4. Capacitor feedback technology	9
Rationale	9
Noise, dynamic range, linearity, and bandwidth	11
Problems	13
How fast is 'fast'?	14
Practical criteria for using capacitor rather than resistor feedback	14
5. Measurement of changes in membrane capacitance	15
6. Series resistance compensation	16
Positive feedback	16
Supercharging	18
7. Patch-pipettes: practical considerations	21
Glasses	21
Pullers	22
Holders	23
8. Conclusion	24
References	24

2. Distinguishing between multiple calcium channel types   27

Martha C. Nowycky

  1. Introduction   27
  2. Major distinguishing features of currently identified calcium channel types   28
    Kinetic properties (voltage- and time-dependence of channel openings and closings)   28
    Permeability   30
    Pharmacology   30
  3. Whole-cell currents   31
    General strategy   31
    Extracellular solutions   32
    Intracellular solutions   33
    Pulse protocols   34
    Permeability   36
    Pharmacology   38
  4. Whole-cell: data analysis   39
    Strategy for isolation   39
    Activation plots   39
    Inactivation plots   39
    Measuring rates of decay   40
  5. Single-channel currents   40
    General strategy   40
    Pipette solution   41
    Bath solution   42
    Conductance (permeability)   42
    Pulse protocols   42
    Pharmacology   42
  6. Single-channel: data analysis   45
    Kinetic analysis   45
    Conductance   45
  7. Summary   45
  8. Supplies   46
    References   46

3. Physiological approaches to the study of glutamate receptors   49

C. E. Jahr and G. L. Westbrook

  1. Introduction   49

2. Experimental preparations	50
3. Pharmacology	52
4. Whole-cell recording	53
Basic issues	54
Analysis of voltage-dependent conductances	55
Open-channel blockers of the NMDA channel	56
Fluctuation (noise) analysis of EAA conductances	56
5. Single-channel recordings	58
N-methyl-D-aspartate channels on cultured neurons	58
Non-NMDA channels	60
Technical issues for recording of single EAA channels	60
6. Concentration clamp techniques	61
Whole-cell recordings	61
Excised patch recording	63
7. Kinetics of synaptic responses	64
Synaptic activation	65
8. Optical measurements	66
Intracellular calcium measurements	66
Chelators of intracellular calcium	68
9. Oocyte expression of glutamate receptors	68
Ion channels	69
Metabotropic glutamate receptor	69
10. Future directions	69
References	70

4. Identifying and characterizing stretch-activated ion channels 75

Christian Erxleben, Joachim Ubl, and Hans-Albert Kolb

1. Introduction	75
Stretch-activated channels: an overview	75
Physiological function of stretch-activated channels	75
2. Methods	76
Recording techniques and analysis of channel fluctuations	76
Hydrostatic pressure and membrane tension	77
3. Theory	78
Pressure and voltage-dependence of the SA-channel open probability	78
4. Comparison of pressure- and voltage-sensitivity of SA-channels in different cell types	80

5. Examples of SA-channels and their proposed physiological function 81
   SA-channels in abdominal stretch-receptor of crayfish 81
   Channel-activation during volume regulation 84

References 90

SECOND-MESSENGER SYSTEMS

5. Identification of G-protein-medicated processes 95

A. C. Dolphin and R. H. Scott

1. Introduction 95
   What is a G protein? 95
   G protein classification 95
   Characteristics of a G-protein-linked system 96
2. Identification of a direct G-protein-mediated process 97
   Summary of criteria for identification of a G-protein-coupled receptor 99
   Summary of criteria for identification of a G-protein-mediated process 99
3. Experimental strategies for the identification of G-protein-mediated signal transduction 99
   Evidence from binding studies for receptor coupling to G proteins 100
   Investigation of G protein GTP-ase activity 100
   Tools to modify G protein activity 101
   Reconstitution studies with exogenous G proteins 106
   Studies of G-protein-mediated processes in intact cells 107
4. Conclusion 109

References 110

6. Modifications to phosphoinositide signalling 115

P. Jeffrey Conn and Karen M. Wilson

1. Introduction 115
2. Buffer composition 118
3. Preparation of slices 118
4. Labelling slices with [^{3}H]inositol 119
   Sources and purification of [^{3}H]inositol 119

Contents

Optimal [³H]inositol and tissue concentrations	119
Duration of incubation with [³H]inositol	120
Pulse-chase labelling with [³H]inositol	122
5. Incubation with lithium and antagonists	122
6. Incubation with agonist	123
Duration of incubation with agonist	123
Incubation with agonist in the absence of LiCl	123
Non-receptor-mediated increases in phosphoinositide turnover	124
7. Extraction of inositol phosphates	125
Chloroform/methanol extraction	125
Trichloroacetic acid extraction	125
Methods for improving inositol phosphate recovery	126
8. Separation of inositol phosphates on Dowex anion-exchange columns	126
Preparation of Dowex columns	126
Elution of inositol phosphates	127
9. Separation of inositol phosphates using HPLC	129
HPLC equipment	129
Elution of inositol phosphates	130
10. Measurement of Ins [1,4,5]P_3 and DAG mass	130
References	132

7. Exogenous kinases and phosphatases as probes of intracellular modulation 135

Michael J. Hubbard and Claude B. Klee

1. Introduction	135
2. General approach	136
3. Detection of phosphoproteins	136
Radiolabelling with ^{32}P	136
Phosphate analysis	139
Characterization of phosphorylation sites	139
4. Identification of protein kinases	139
Assay of protein kinase activity	140
Primary structure of substrate	143
Calmodulin kinase II	143
5. Identification of protein phosphatases	145
Assay of protein phosphatase activity	146
Calcineurin	149
6. Testing for effects of phosphorylation on function	150

7. Establishing a physiological role	151
References	154

MOLECULAR BIOLOGICAL APPROACHES

8. The expression of neurotransmitter receptors and ion channels in *Xenopus* oocytes — 161

John P. Leonard and Terry P. Snutch

1. Introduction	161
2. RNA synthesis and purification	162
General considerations	162
RNA isolation	164
Synthetic RNA	166
RNA size fractionation	167
3. Preparation and injection of oocytes	171
Xenopus frogs and oocytes	171
Injection and culture of oocytes	174
4. Electrophysiology of channels expressed in oocytes	175
Two-electrode voltage clamping	175
Patch clamping	177
Big patches	177
Single-channel recording	178
Isolation of exogenous currents	178
References	180

9. Molecular approaches to the structure and function of the GABA$_A$ receptors — 183

F. Anne Stephenson and Michael J. Duggan

1. Introduction	183
2. Synthetic peptides as antigens	185
The choice of synthetic peptide	185
The method of covalent coupling of peptide and carrier	188
3. Antibody production and antibody screening methods	191
The ELISA and immunoblotting screening methods	192
The soluble immunoprecipitation assay	199
4. Conclusions	202
References	203

10. In situ hybridization with synthetic DNA probes 205

William Wisden, Brian J. Morris, and Stephen P. Hunt

   1. Introduction: mRNA hybridization using synthetic oligodeoxyribonucleotide probes 205
      Methods included in this chapter 206
   2. Oligonucleotide probes 206
      Probe synthesis 206
      Purification of probes 208
   3. Labelling of probes 209
      The terminal transferase reaction 209
      Analysis of labelled probes on acrylamide gels 211
   4. In situ hybridization 211
      Materials and reagents 213
      Preparation and fixation of tissue 215
      Hybridization of probes to tissue 217
      Autoradiography of sections 219
   5. Controls for in situ hybridization 219
   6. Prospects 223
      References 223

Contents of Cellular neurobiology: a practical approach 227

Index 229

Contributors

P. JEFFREY CONN
Department of Pharmacology, Emory University School of Medicine, Atlanta, Georgia 30322, USA.

A. C. DOLPHIN
Department of Pharmacology, Royal Free Hospital School of Medicine, Rowland Hill St., London NW3 2PF, UK.

MICHAEL J. DUGGAN
Department of Pharmaceutical Chemistry, School of Pharmacy, 23–29 Brunswick Square, London WC1N 1AX, UK.

CHRISTIAN ERXLEBEN
Fakultät für Biologie, Universität Konstanz, Postfach 5560, 7750 Konstanz, Germany.

ALAN S. FINKEL
Axon Instruments, Inc., 1101 Chess Drive, Foster City, CA 94404, USA.

MICHAEL J. HUBBARD
Department of Biochemistry, University of Otago, Dunedin, New Zealand.

STEPHEN P. HUNT
MRC Molecular Neurobiology Unit, University of Cambridge Medical School, Hills Road, Cambridge CB2 2QH, UK.

C. E. JAHR
Vollum Institute L474, Oregon Health Sciences University, 3181 Sam Jackson Park Road, Portland, OR 97201, USA.

CLAUDE B. KLEE
Department of Health and Human Services, National Institutes of Health, Bldg. 37 Room 4C–06, Bethesda, MD 20892, USA.

H. A. KOLB
Fakultät für Biologie, Universität Konstanz, Postfach 5560, 7750 Konstanz, Germany.

JOHN P. LEONARD
Department of Biological Sciences, University of Illinois at Chicago, Box 4348, Chicago, IL 60680, USA.

BRIAN J. MORRIS
MRC Molecular Neurobiology Unit, University of Cambridge Medical School, Hills Road, Cambridge CB2 2QH, UK.

Contributors

MARTHA C. NOWYCKY
Department of Anatomy, Pennsylvania Medical College, 3200 Henry Ave., Philadelphia, PA 19129, USA.

R. H. SCOTT
Department of Physiology, St. George's Hospital Medical School, London, SW17 0RE, UK.

TERRY P. SNUTCH
Biotechnology Laboratory, RM 237–6174 University Boulevard, University of British Columbia, Vancouver, BC, Canada V6T 1WS.

F. ANNE STEPHENSON
Department of Pharmaceutical Chemistry, School of Pharmacy, 29–39 Brunswick Square, London, WC1N 1AX, UK.

JOACHIM UBL
Fakultät für Biologie, Universität Konstanz, Postfach 5560, 7750 Konstanz, Germany.

G. L. WESTBROOK
Vollum Inst. L474, Oregon Health Sciences University, 3181 SW Sam Jackson Park Road, Portland, OR 97201, USA.

KAREN M. WILSON
Department of Pharmacology, Emory University School of Medicine, Atlanta, Georgia 30322, USA.

WILLIAM WISDEN
Laboratory of Molecular Neuroendocrinology, ZMBH, University of Heidelberg, Im Neuenheimer Feld 282, 6900 Heidelberg, Germany.

Abbreviations

ACh	Acetylcholine
AChR	acetylcholine receptor
ACSF	artificial cerebrospinal fluid
ADP	after-depolarization
AHP	after-hyperpolarization
ARF	ADP-ribosylation factor
ATP	adenosine triphosphate
BAPTA	1,2 bis (2-amino phenoxy)ethane-N,N,N',N'-tetra acetic acid
BDNF	brain-derived neurotrophic factor
BK	bradykinin
BSA	bovine serum albumin
BSS	balanced salt solution
CAM	cell-adhesion molecules
cAMP	cyclic adenosine monophosphate
cGMP	cyclic guanylate monophosphate
cGRP	calcitonin-gene-related peptide
CM	conditioned medium
CNS	central nervous system
CNTF	ciliary neurotrophic factor
DAG	diacylglycerol
DEAE	diethylaminoethyl
DEPC	diethylpyrocarbonate
DHP	dihydropyridine
DM	defined medium
DMSO	dimethyl sulphoxide
DRG	dorsal root ganglion
DSP	digital signal processors
DTT	dithiothreitol
EDTA	ethylene diamine tetraacetate
EGTA	ethyleneglycolbis-(β-aminoethyl)ethane-N,N,N',N'-tetra acetic acid
EMEM	Eagle's minimum essential medium
ELISA	enzyme-linked immunoabsorbent assay
EPSP	excitatory post-synaptic potential
EPSC	excitatory post-synaptic current
FCS	fetal calf serum
GABA	gamma-aminobutyric acid
GDP	guanosine diphosphate
GTP	guanosine triphosphate
HPLC	high performance chromatography

Abbreviations

HRP	horseradish peroxidase
IBMX	isobutylmethylxanthine
IPSC	inhibitory post-synaptic current
IPSP	inhibitory post-synaptic potential
KLH	keyhole limpet hyamocyanin
LTP	long-term potentiation
NAD	nicotinamide adenine dinucleotide
NGF	nerve growth factor
NMDA	N-methyl-D-aspartate
NmDG	N-methyl-D-glucamine
PBS	phosphate buffered saline
PNS	peripheral nervous system
PLC	phospholipase C
PSPs	post-synaptic potentials
SDS	sodium dodecyl-sulphate
SLT	superlateral radular tensor
SSC	standard saline citrate
TBA	tetrabutylammonium
TEA	tetraethylammonium
TLC	thin-layer chromatography
TTX	tetrodotoxin
UV	ultraviolet

IONIC CHANNEL RECORDINGS

1

Progress in instrumentation technology for recording from single channels and small cells

ALAN S. FINKEL

1. Introduction

The development of the single-channel patch clamp marked a stunning evolution in our capacity to observe and understand membrane conductance processes. For the first time, the dynamic activity of individual membrane molecules could be observed. This was rare luck: there are very few disciplines in which this is possible. From the beginning, there have been arguments about the physiological significance of currents recorded from membranes that have been stretched across the mouth of the recording pipette and whose inner surfaces have been dialysed, but notwithstanding, an enormous amount of valuable data has been gathered using this technique.

The technology emerged relatively recently. Intracellular micropipettes began to be used widely from the early 1950s, whereas the patch-clamp technique only entered widespread use in the early 1980s.

The very first single-channel recording was made in 1976 by Neher and Sakmann (1), but the technique did not become popular until major advances in instrumentation technology were made. Progress was not limited to single-channel recording: rapid enhancements in the techniques for recording whole-cell currents from small cells emerged simultaneously with the development of the techniques for recording single-channel currents. Although the patch-clamp techniques are now used routinely, advances are still being developed that extend the limits of what can be resolved. Some of these technologies are discussed in this chapter.

2. Background

2.1 What is a patch clamp?

The patch clamp is a special case of the voltage clamp. In the voltage-clamp technique, the membrane potential is held constant while the current flowing through the membrane is measured. Typically, the investigator has no inherent interest in the membrane *current*—he or she is interested in the membrane *conductance*, since conductance is directly proportional to the ion-channel activity. Membrane current is measured because there is no direct way to measure membrane conductance. By holding the membrane potential constant (or, at the very least, constant after a rapid step), the investigator ensures that the current is linearly proportional to the conductance that he or she is interested in.

Microelectrode voltage-clamp techniques for whole-cell current measurement traditionally assign the role of potential measurement and current passing to two different intracellular electrodes. This is done using the conventional two-electrode voltage-clamp technique, or it is simulated by time division multiplexing using the switched single-electrode voltage-clamp technique. This separation of the microelectrode roles avoids the introduction of errors in the measurement due to unknown and unstable voltage drops across the series resistance of the current-passing electrode.

Instead of using sharp microelectrodes to puncture the membrane and penetrate the cell like in traditional voltage clamping, patch clamping uses a heat-polished pipette of about 1 μm or more in diameter that is sealed to a 'patch' of membrane. The same pipette is used continuously for both current passing and voltage recording.

Two major classes of patch clamp are common. These are the whole-cell patch clamp and the single-channel patch clamp. A third class, called the loose patch clamp is also used, but not widely.

In the whole-cell patch clamp, the membrane at the tip of the pipette is ruptured. The electrolyte in the pipette is thus in electrical continuity with the interior of the cell and the total membrane current is recorded. In this case, the current through the series resistance of the pipette and the residual resistance of the ruptured patch is often sufficiently large to introduce significant voltage errors. Techniques exist to compensate for these errors but the compensation is never perfect. To get a feeling for the magnitude of the errors, assume that the maximum compensation is 80%, beyond which the system oscillates and destroys the cell. Further, assume that the access resistance (R_a; the sum of the pipette resistance and the residual resistance of the ruptured patch) is 5 MΩ. After compensation, the effective value of R_a ($R_{a,eff}$) is just 1 MΩ. In this case, a 10 nA current would result in a 50 mV uncompensated voltage error, reduced to 10 mV by the compensation. Clearly, the whole-cell patch-clamp technique cannot be used to record large currents and even for modest whole-cell currents care must be taken to compensate for the series resistance and then to correctly interpret the residual error.

The temporal resolution of the whole-cell patch clamp is also affected by $R_{a,eff}$. The time-constant for resolving currents is the product of $R_{a,eff}$ (assuming that the membrane resistance is much greater) and the membrane capacitance ($\tau = R_{a,eff}C_m$). Thus the technique is also limited to small cells where this product is small enough for the desired time-resolution to be achieved. *Figure 1* illustrates the voltage and temporal errors caused by the presence of R_a.

In the single-channel patch clamp, the membrane at the tip of the pipette is preserved. The current recorded is just that current which flows through the patch of membrane enclosed by the tip of the pipette. Since the area of this membrane is very small, there is a good chance that just one or a small number of ion channels may be in the membrane patch. Individual ion-

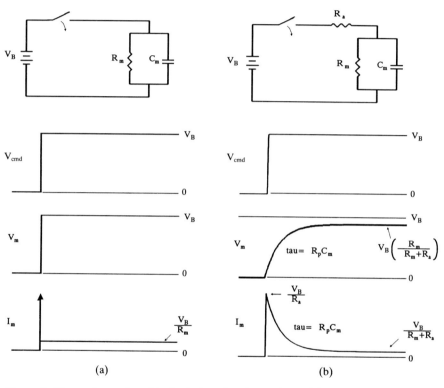

Figure 1. Comparison of an ideal whole-cell voltage clamp (a) with a real whole-cell voltage clamp (b). The *ideal* clamp (a) simply consists of a battery, a switch, a wire and the cell. In the *real* clamp, there is an access resistance R_a (which would be replaced by $R_{a,eff}$ if series-resistance compensation were used). In the ideal case the membrane charges instantly to the battery voltage. There is an impulse of current injecting a charge $Q = C_m V_B$; the steady-state current is V_B/R_m. In the real case (b), the charging current is limited by R_a. The initial amplitude is limited to V_B/R_a; there is a steady-state current equal to $V_B/(R_a + R_m)$. The time-constant for charging the membrane capacitance is $\tau = R_p C_m$ (where R_p is the value of R_a and R_m in parallel). The injected charge has two components, the transient component and the steady-state charge. The transient charge is $Q = C_m V_B R_m/(R_m + R_a)$. The steady-state component is simply the steady-state current integrated for the observation time.

channel currents can thus be recorded. In this case, the current through the series resistance of the pipette is very small, perhaps a few picoamps through a series resistance of just 10 MΩ. The resulting error voltage is just a few tens of μV and is always ignored.

2.2 The importance of a good seal

When a heat-polished pipette is pressed against the cell membrane, the interior of the pipette is isolated from the extracellular solution by the *seal* that is formed. If the resistance of this seal is infinite, no current can leak across this seal.

The leakage current through the seal resistance has a crucial bearing on the quality of the patch current recording. In the first place, depending on the size of the seal resistance and the voltage difference between the inside and the outside of the pipette, some fraction of the current through the patch of membrane will leak out through the seal and will thus not be measured. The lower the seal resistance, the larger the fraction of undetected current.

In the second place, thermal movement of charges in the conducting pathways of the seal constitute the major source of noise in the recording unless the seal resistance is very high (several gigaohms or more). A high seal resistance is a prerequisite of low-noise recordings.

2.3 A brief history

Heat-polished pipettes were initially used to record the current through rather large areas of membrane using pipettes of about 10 μm in diameter. These recordings were useful for studying the membrane impedance (2) and for localizing the active currents through the membrane (3, 4) but it was not possible to resolve individual channel currents.

The first recordings of currents through single biological channels were made in 1976 by Neher and Sakmann using denervated frog muscle fibres (1). Their success in achieving this milestone was crucially dependent on an enzymatic treatment that they developed to remove the basement membrane and the connective tissue covering the cell surface. Once these tissues were removed, the contact of the pipette tip with the membrane was better and therefore the seal resistance was higher (about 20–50 MΩ) than had previously been achieved. However, although Neher and Sakmann were able to resolve single-channel currents and, for the first time, show that these were rectangular pulses of fixed amplitude but random duration, the background noise was still quite high (3 pA peak-to-peak in a 200 Hz bandwidth). Therefore, the technique was only applicable to instances where the channel currents were large.

The explosion of research using the single-channel patch technique came later as a result of the discovery of the method to form giga-seals. Giga-seals are seals in which the seal resistance is so high (10–100 GΩ) that the leakage current through the seal resistance is, in a practical sense, irrelevant. This discovery was made by Neher in 1980. He discovered that if very clean pipettes were used and that if they were only used once, giga-seals would form almost routinely, especially if gentle suction was applied. This discovery was made possible by Neher's development of a modified pipette puller that allowed him to mass produce pipettes which he then used one after the other.

Once giga-seals were formed and the noise due to the leakage current was virtually eliminated, other sources of noise emerged as the dominant limitations in the resolution of the recording technique. These were the noise of the electronics, the glass, the input capacitance and the feedback resistor. Work in Neher's and Sakmann's laboratories lead to the design of lower-noise current-to-voltage amplifiers and the use of higher value resistors as well as the selection of glasses having better dielectric properties. Sylgard was used to lower the overall input capacitance by increasing the effective thickness of the wall of the pipette and by preventing the solution from creeping up the outer wall of the pipette. The details of the procedures and equipment were published in 1981 in a seminal paper produced by the members of Neher's and Sakmann's laboratories (5). This paper is still the standard reference for the single-channel patch-clamp technique, supplemented by a book published in 1983 that described the single-channel recording technique in even greater detail (6). An interesting account of the discovery process was published by Sigworth in 1986 (7).

3. Resistor feedback technology

3.1 Description of the current-to-voltage headstage

The basic concept behind the design of the patch-clamp electronics is very simple (*Figure 2*). A sensitive current-to-voltage converter is fabricated using a high-megaohm resistor (R_f) and an operational amplifier (op amp, A1). The pipette is connected to the negative input and the command voltage is connected to the positive input. Since the op amp has extremely high gain, the potential at the negative input is forced to follow the potential at the positive input. All current flowing in the micropipette also flows through R_f. This current is proportional to the voltage across R_f, which is measured at the output of the differential amplifier (A2).

In principle, the patch clamp is equivalent to a conventional two-electrode voltage clamp in which the output circuit is connected back to the input electrode. In practice, the patch clamp is better behaved. First, since all of the gain is in a single op amp stage, the system is simpler. Second, the stray capacitance across the feedback resistor guarantees stability. Third, because the gain of the op amp is so high, the difference between the command potential and the potential of the micropipette is negligible. Note that the potential that is controlled is the potential at the shaft of the micropipette, not the potential at its tip (as desired).

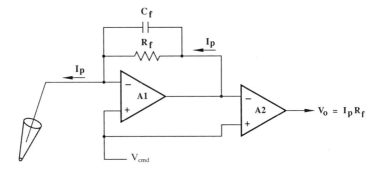

Figure 2. Simplified schematic of the resistor feedback patch-clamp headstage. Op amp A1 and its feedback resistor, R_f, constitute a current-to-voltage converter. Because of its high gain, the op amp forces the potential at the negative input to follow the command potential (V_{cmd}) at the positive input. The pipette current is identical to the current in R_f (assuming a small and constant bias current). The current in R_f generates a voltage across it that is measured in the differential amplifier A2. C_f represents the stray capacitance across R_f. In practice, C_f consists of many resistor-capacitor components that make the frequency response of R_f very complicated.

3.2 Problems

The special demands of patch clamping lead to complex considerations in the design of a good headstage.

(a) Integrated circuit op amps do not have the required combination of ultra-low noise and sub-picoamp bias currents. Thus the op amp has to be made from a composite of low-noise field-effect transistors (FET) such as the U430 type (Siliconix, Santa Clara, California, USA) and conventional op amps (8).

(b) The minimum theoretically achievable current noise is smaller for larger R_f values, therefore it is important to choose large values. 50 GΩ is the largest value that is typically used, since it still allows a reasonable maximum current of more than 200 pA to be measured. Unfortunately, for reasons that are not well understood, these high-value resistors are several times noisier than predicted by thermal-noise theory. The noise of these resistors cannot be predicted. Various brands of resistors must be tested until the best one is empirically found.

(c) The inherent bandwidth of a 50 GΩ resistor is limited by the stray capacitance across it (C_f in *Figure 2*). For example, a 50 GΩ resistor with 0.1 pF of stray capacitance has a 5 msec time-constant, corresponding to a bandwidth of only 32 Hz. This poor time-resolution is unacceptable for measuring ionic currents, so the high-frequency components of the headstage output signal must be boosted. This is typically achieved by an analog frequency compensation circuit. The design of this circuit is complicated by the fact that R_f cannot be considered as an ideal capacitor in parallel with an ideal resistor, thus a simple high-pass filter cannot be used. Complex circuits consisting of up to four poles and three zeros in the transfer function are commonly used (8). The placement of the poles and zeros must be carefully adjusted for the particular resistor. An alternative to the analogue boost circuit would be to use a digital correction scheme. To date, this has not been practical to achieve, because many millions of floating point calculations would be needed per second. However, the recent emergence of high-speed digital signal processors (DSPs) should make this approach possible. It would have the advantages of being easier to set up and of being noise-free (assuming the digitization is high resolution—16 bits or more—and that measures are taken to match the range of the headstage output to the range of the analogue-to-digital converter).

(d) The current required to charge the input capacitance during a step-voltage command can easily exceed the maximum current that can be passed by R_f from the typical ±13 V swing of the op amp. For example, to linearly charge 5 pF of input capacitance to 100 mV in 10 μsec would require a charging current of 50 nA. This is well beyond the 260 pA that can be passed by a 50 GΩ resistor driven from 13 V. Thus, special circuits are added to inject the required charging current through a capacitor (9).

(e) The best high-value resistors seem to be available only in chip form. This means that the headstage electronics have to be manufactured in a hybrid. This manufacturing technique has some advantages. First, it means that the U430 input FETs and other FETs typically used to switch between different values of R_f can be used in chip form. This is to be preferred, since the sealing glasses used in packaged transistors typically have a leakage resistance that increases the noise. Second, it means that the sensitive input components can be maintained in a hermetically sealed environment that decreases the rate at which they age due to environmental contamination.

4. Capacitor feedback technology

4.1 Rationale

A very new technology in single-channel recording is the capacitor-feedback technique, also known as the integrating headstage technique. While this technique was talked about for many years, it was first attempted and described in 1985 (10). Practical implementations did not become commercially available until 1989 (IHS-1 and Axopatch 200 (from Axon Instruments Inc., Foster City, California, USA), 3900 (from Dagan Corporation, Minneapolis, Minnesota, USA).

In this technique, the feedback resistor is replaced by a capacitor (C_f, *Figure 3*). When a capacitor is used instead of a resistor, the headstage acts as an integrator. That is, for a constant input current, the output of the

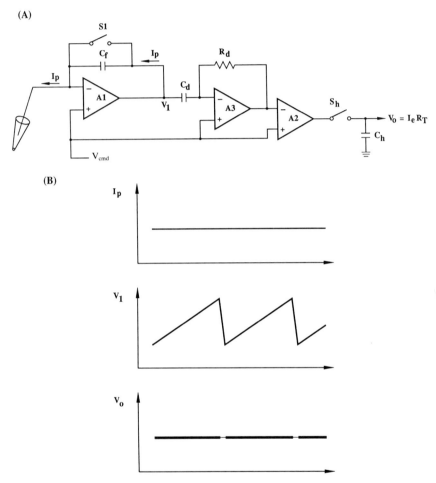

Figure 3. (A) Simplified schematic of the capacitor-feedback patch-clamp headstage. As in the resistor-feedback headstage, op amp A1 and its feedback element, C_f, constitute a current-to-voltage converter. The pipette current is identical to the current in C_f. The magnitude of the current is proportional to the slope of the voltage across C_f. This slope is measured by the differentiator, C_d, R_d, and A3. Whenever the value of V_1 exceeds an absolute value of 10 V, switch S1 closes to reset the capacitor. During the reset time, the output of the differentiator undergoes a large transient (not shown). The sample-and-hold circuit at the output (S_h and C_h) is used to blank this transient from the output. (B) If the pipette current is constant, the identical constant current in C_f generates a voltage ramp on the V_1 output of A1. This constant current is measured by the differentiator. The thin portions in the output trace represent the periods when the output is blanked.

headstage is a ramp. The slope of the ramp is proportional to the current. To recover a voltage that is linearly proportional to the input current, the integrator must be followed by a differentiator (A3).

Compared with resistor-feedback picoamp current-to-voltage converters (RIV), capacitor-feedback current-to-voltage converters (CIV) exhibit less noise, increased dynamic range, wider bandwidth, and improved linearity.

4.2 Noise, dynamic range, linearity, and bandwidth

(a) Two factors contribute to the lower noise of the CIV. First, the theoretically predicted thermal noise of resistors is not present in capacitors. (There is an equivalent resistor noise source due to the differentiator, but with careful design this can be made negligible.) Second, capacitors are commercially available that are free of the excess noise sources that plague high-gigaohm resistors. In practical implementations, the noise of a CIV is less than 0.2 pA r.m.s. in a 10 kHz bandwidth compared with 0.3 pA for an RIV.

The noise benefit of the CIV is eliminated if the added capacitance at the input of the headstage is large (say 10 pF or more). This is because the voltage noise of the input FETs causes current noise to be injected into the input capacitance. If the input capacitance is large, this source of noise becomes dominant.

The lower noise of the CIV can only be realized in situations where all other noise sources are minimized. That is, low-noise glass and holders should be used, the pipettes should be sylgard-coated, and high-resistance seals are required. When these precautions are taken, the reduced noise of the CIV can be beneficial in real patches, not just in theory (see *Figure 4A*).

(b) The capacitor-feedback headstage has an equivalent transfer resistance (R_T) given by

$$R_T = R_d(C_d/C_f).$$

Because the noise is theoretically independent of the C_f value and the gain of the differentiator, R_T may be kept quite low, say 100 MΩ. At this gain, the maximum current that can be recorded is 500 times greater than for an RIV using a 50 GΩ feedback resistor. Thus, the CIV potentially has a vastly improved dynamic range.

The main area where this increased dynamic range is useful is the recording of single-channel currents from bilayers, since large currents have to be passed during a voltage step to charge the bilayer capacitance. In order to charge the large bilayer capacitance using a resistor-feedback headstage, the headstage has to be specially designed so that the feedback resistor can be changed by a control pulse from a low value (e.g. 50 MΩ) for charging the membrane capacitance up to the high value used for single-channel recording (50 GΩ). Headstages with this capability are commercially available (CV-4B, Axon Instruments Inc., Foster City, California, USA).

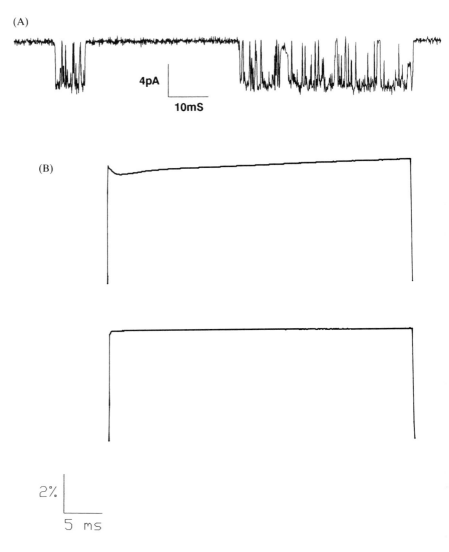

Figure 4. (A) Example of a single-channel recording using a capacitor-feedback headstage. Currents shown are from a stretch-activated potassium channel from rabbit corneal epithelium. Recorded by James L. Rae using an Axon Instruments IHS-1 headstage. Bandwidth: 5 kHz. (B) The upper panel shows the step response of a resistor-feedback headstage after trimming with 4 poles and 3 zeros. The response is almost square, but there are some small undulations amounting to about 2% of the step response. Note that the undulations have been exaggerated for clarity by only plotting the final 6% of the step response. The lower panel shows the step response of a capacitor-feedback headstage without any trimming. There are no undulations. Bandwidth: 2 kHz. The lower panel shows the step response of a capacitor-feedback headstage without any trimming. There are no undulations. Bandwidth: 2 kHz. Pulse height 100% approximately 100 pA; pulse width 30 msec.

(c) In the section on resistor feedback technology, it was pointed out that in order to achieve an acceptable bandwidth of about 20 kHz for single-channel recording, complex boost circuitry has to be used to correct the frequency response. Even if the boost circuitry has as many as four poles and three zeros, the frequency response of R_f is not perfectly corrected. The imperfections of the correction are most easily seen when observing the step response of the headstage. Typically, there will be overshoots and ripples that can amount to as much as 2% of the response. With a CIV, because of the excellence of the available capacitors, the step response is inherently square (see *Figure 4B*).

(d) The bandwidth of a CIV can be very wide. It is maximized by using a small value of C_f, and a small value of R_d. For practical values ($C_f = 1$ pF and $R_d = 100$ kΩ), the bandwidth is of the order of 70–80 kHz. This is considerably better than the 20 kHz of a good RIV using a 50 GΩ feedback resistor.

Whether it is practical to use this increased bandwidth or not is another matter. The noise of both the RIV and the CIV headstages increases dramatically with frequency. While the noise in a 10 kHz bandwidth is 0.2 pA r.m.s. (about 1.6 pA peak-to-peak), the noise in a 50 kHz bandwidth is about 1.8 pA r.m.s. (about 14 pA peak-to-peak). Thus the single-channel currents would have to be extremely large to be distinguished from the noise.

4.3 Problems

There are two major problems that make the capacitor-feedback technology difficult to implement.

4.3.1 Resets

The first problem is that after a sustained net DC input current, the integrator voltage ramps towards one of the power-supply limits. When this happens, the integrator must be reset by discharging the capacitor. The frequency of the resets depends on the size of C_f and the average input current. For example, if C_f is 1 pF and the input current is 2 pA, the output of the integrator will ramp at 2 V/sec. If the reset circuitry is set to act when the integrator voltage exceeds 10 V, resets will occur every 5 sec.

The reset itself takes of the order of 50 μsec. During the reset, sample-and-hold circuits are used to maintain the current output at its value immediately prior to the start of the reset. If it is tolerable to lose 0.1% of the data, resets as frequent as every 50 msec would be acceptable. In the above example, this corresponds to an input current of 200 pA. For single-channel recording, this current is more than adequate.

For whole-cell recordings and their larger currents, the resets would be too frequent unless the value of C_f was increased. In practice, it is hardly worth it. It is simpler to switch to a modest-sized resistor (e.g. 500 MΩ) and avoid the problem of resets altogether. Reverting to resistor-feedback methods for whole-cell recordings does not represent a set-back in terms of noise because the whole-cell currents are often quite large, and because the excess noise of 500 MΩ resistors is not comparatively as bad as the excess noise of 50 GΩ resistors.

During the measurement of voltage-activated single-channel currents, the resets can be made totally irrelevant. This is done by using a control pulse to force a reset immediately prior to the voltage step. This guarantees that C_f is in its discharged state at the onset of the voltage step and is, therefore, highly unlikely to require resetting during the voltage step.

4.3.2 Transients

The second problem is that during resets, transients are injected into the headstage input by the reset circuitry. If left uncompensated, these would cause unwanted and possibly damaging current pulses to be injected down the micropipette into the patch. Special compensation circuitry needs to be included in the design to exactly balance out these injected currents.

There are other transient problems that occur during and after the reset. These result from dielectric absorption in the differentiator and feedback capacitors and other ill-defined transients that work their way into the system. Compensation circuitry must also be included to compensate for these transients.

Overall, the need to reset and to compensate for the many transients that occur during reset make the design of an integrating headstage extremely difficult. However, in a well-designed system, the resent transients measured at the output will be less than 1 pA measured in a 1 kHz bandwidth.

4.4 How fast is 'fast'?

The speed of the observed transitions of single-channel currents is the same as the response time of the electronics. Whenever recordings are made at wider bandwidths, the observed transition rates become shorter. It is not clear that we will ever be able to resolve the actual transition times using electrical measurements (perhaps optical techniques will emerge to do the job), but the discovery of new lower limits to the transition times will be made possible by the new capacitor-feedback patch clamps, assuming that large channels are studied. (Large channels will be necessary so that the channel currents will be visible against the large background noise associated with wide recording bandwidths.)

4.5 Practical criteria for using capacitor rather than resistor feedback

The choice between a resistor-feedback or a capacitor-feedback patch clamp is a difficult one. A number of considerations must be considered. The main advantage of the resistor-feedback patch clamps is that there are no reset transients. The technology is well-proven and well-understood. In many experiments, the ultimate in noise, bandwidth and linearity are not required and in these cases the simplicity of a resistor-feedback patch clamp might be attractive. In other cases, the reduction in background noise afforded by a well-designed capacitor-feedback patch clamp might mean the difference between resolving the closings in a burst or not. If noise or bandwidth is the limiting factor in the experiment there is no question but to use a capacitor-feedback patch clamp.

A capacitor-feedback patch clamp is also desirable for bilayer patch clamping. This is not because of any supposed noise benefit; as it happens, there is no noise benefit in a typical bilayer experiment, because the dominant source of noise for either type of patch clamp is related to the product of the noise of the input FET and the bilayer capacitance. Rather, the capacitor-feedback patch clamp is desirable because of its ability to rapidly charge the bilayer membrane during voltage steps.

5. Measurement of changes in membrane capacitance

The measurement of minute changes in membrane capacitance such as those occurring during exocytosis (11, 12) was made practical by the whole-cell patch clamp technique. Two methods can be used. The simplest, and the most traditional, is to use a least-squares iteration technique to find the time-constant of the current response to a voltage step (13). By assuming a simple cell model, the membrane capacitance can be easily deduced. This technique is relatively easy to apply, but it has a resolution no better than 100–200 fF. This is insufficient to resolve the 10–20 fF capacitance increases occurring during fusion of single granules with the membrane (11).

A much more sensitive technique involves the use of a lock-in amplifier (also known as a phase detector). In this technique, a sinusoidal command voltage is applied to the cell. The magnitude of this voltage must be small enough so that the cell's properties are essentially linear. Two outputs proportional to the sinusoidal current response at two orthogonal frequencies are measured. It can be shown that the magnitudes of these two responses are a function of the access resistance, the membrane resistance and the cell capacitance. With a third measurement of the DC current, and assuming that the reversal potential of the DC currents is known, there is sufficient information to calculate the value of all three of the parameters (13). The resolution of this technique can be as good as 1 fF, with measurements being made about once per second.

The lock-in amplifier technique requires special equipment that is not widely available. Joshi and Fernandez (14) have recently described a similar technique where all of the phase-sensitive measurements and sinusoidal stimulations are performed by software. This has the great advantage of not requiring any special equipment other than the patch clamp itself. This software technique should make it easier for investigators to perform sensitive capacitance measurements.

6. Series resistance compensation

As discussed in Section 2.1, in the ideal experiment the resistance of the patch micropipette in whole-cell experiments would be zero. In this case, the time resolution for measuring membrane currents and changing the membrane voltage would be limited only by the speed of the electronics (typically just a few microseconds).

6.1 Positive feedback

Series resistance compensation using positive feedback is an attempt to achieve this ideal electronically. This technique is also called 'correction'. It is the same technique that has been widely used in conventional two-electrode voltage clamps (15, 16). Basically, a signal proportional to the measured current is used to increase the command potential. This increased command potential compensates for the potential drop across the micropipette (see *Figure 5*). The amount of compensation achievable is limited by two considerations. First, as the compensation level approaches 100%, the increase in the command potential hyperbolically approaches infinity. For example, at 90% compensation, the command potential is transiently increased by a factor of ten ($V_{cmd}/(1 - \alpha)$, where α is the fractional compensation). Thus, at large compensation levels the electronic circuits approach saturation. Second, the current feedback is positive. The stability of the circuit is degraded by the positive feedback and at 100% compensation

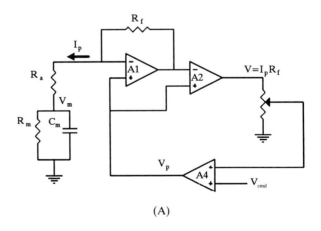

(A)

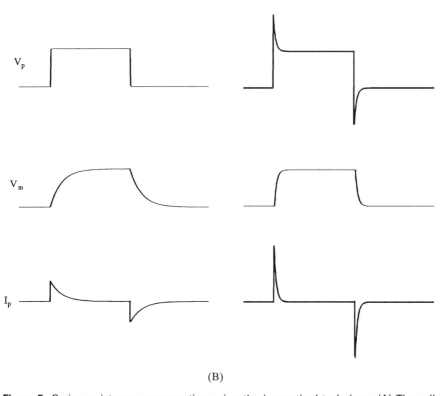

Figure 5. Series resistance compensation using the 'correction' technique. (A) The cell (R_m and C_m), the access resistance (R_a), and the headstage electronics (A1, R_f, and A2) are as described previously. Two extra voltages are indicated: V_m is the true membrane potential. This can only be measured by using a separate recording electrode: V_p is the sum of all command potentials. It is equal to the voltage at the shaft of the pipette. A potentiometer at the output of the headstage is used to determine a scaled signal proportional to the current (I_p). This is summed with the command signal (V_{cmd}) to form V_p. (B) The traces show the responses recorded with an artificial cell (R_m = 500 MΩ, C_m = 33 pF, R_a = 10 MΩ). In the left-hand panel, no series-resistance compensation is used. V_p is equal to V_{cmd}. The time-constant for V_m and I_p is $\tau \approx R_a C_m$. In the right-hand panel, 80% series-resistance compensation is used. The positive feedback action superimposes an exponential 'blip' on V_p that causes the membrane potential to charge to its final value more rapidly. The time-constant for V_m and I_p is $\tau \approx 0.2 R_a C_m$.

the circuit becomes an oscillator. In practice, the oscillation point is lower than 100% because of non-ideal phase shifts in the micropipette and the cell membrane.

The first problem, saturation of the electronics, could in principle be reduced by using high-voltage (e.g. ±120 V) op amps, but this approach has not been pursued because these types of op amp have more noise and worse drift than good conventional op amps. Instead, it is normal to use an injection capacitor connected to the pipette to pass the current required to charge the membrane capacitance (9). This capacitively supplied current unburdens the feedback resistor (R_f) from the necessity of passing the charging current into the cell.

The second problem, stability, has been partially reduced in recent years by adding a variable low-pass filter in the current-feedback loop [e.g. the 'lag' control of the Axopatch-1D (Axon Instruments) or the 'slow' setting of the EPC-7 (List Electronic, Darmstadt-Eberstadt, FRG)]. By empirically setting the low-pass filter cut-off frequency, large percentage compensations can be used, but these only apply to the currents at bandwidths below that of the filter cut-off. Thus the DC, low, and mid-frequency series resistance errors can be substantially reduced while the high-frequency errors remain large.

6.2 Supercharging

Another technique for speeding up the step voltage change is the supercharging technique (17, 18), also known as 'prediction'. In contrast to the correction method of series-resistance compensation, this technique is an open-loop method. That is, there is no feedback and there is no risk of oscillations.

Supercharging is accomplished by adding a brief 'charging' pulse at the start and the end of the command voltage pulse. This means that initially the membrane charges towards a larger final value than expected (see *Figure 6A*).

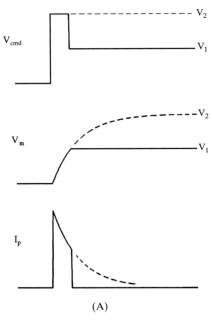

(A)

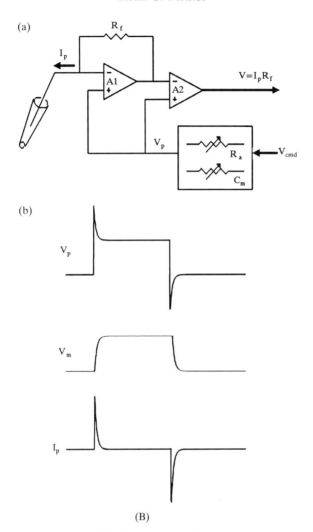

Figure 6. Supercharging (prediction) using two different methods of estimating the optimal V_{cmd}.
(6A) Empirical estimation of V_{cmd}. An initial rectangular overshoot is added to V_{cmd}. This causes the membrane potential to initially charge towards the amplitude of the rectangular overshoot (V_2). At the time that V_m transits through the intended final value (V_1), the overshoot is terminated and V_m remains at V_1. The height and duration of the overshoot in V_{cmd} is determined empirically by minimizing the width of the current transient. In all cases, the charging time-constant for V_m and I_p is the same.
(6B) Automatic estimation of V_p. (a) The circuitry for estimating R_a and C_m is complex and is not shown. The principles are discussed in ref. 9. The two potentiometers marked R_a and C_m are meant to represent this circuitry. (b) The overshoot added to V_{cmd} in this case is an exponential. It has the time constant of the supercharged cell. The charging time constants for V_p, V_m and I_p are the same. This figure shows 80% compensation in the same model cell used in *Figure 6*.

The charging pulse amplitude or time is adjusted empirically by the investigator so that the membrane potential does not overshoot. Since the membrane potential is not directly observable by the user, this is accomplished by adjusting the controls until the current capacitance transient is as brief as possible. Once the optimum setting for one step size has been found, the size of the supercharging pulse for all other step sizes can be calculated by the computer.

Recently, another technique for generating the supercharging current was introduced (R.A. Levis, personal communication, 1989). It is based on having complete knowledge of the R_a and C_m values. Fortunately, this information is generally available on modern whole-cell patch-clamp amplifiers as a by-product of the process of setting the parameters of the circuit that drives the injection capacitor (used to supply the membrane capacitance charging current). Once these parameters are determined it is possible to automatically boost changes in the command voltage to supercharge the cell (Figure 6B). This technique has the significant advantage that no empirical determination needs to be made and that it works with any shape command voltage without requiring a computer to calculate the supercharging pulse.

Supercharging is appropriate for a specific type of experiment, but it is inappropriate in others and its use should be carefully considered. The appropriate application is the measurement of ionic currents that activate during the time that the membrane potential is normally changing. If supercharging is applied, it might be possible to clamp the potential to its final value before the ionic current is substantially activated.

There are two significant shortcomings that the investigator should be aware of when using the supercharging technique:

(a) The technique does not correct for the voltage error that occurs when current flows through the series resistance of the pipette.

(b) The dynamic response of the circuit is not improved. That is, there is no improvement in the speed with which membrane current changes (and hence conductance changes) can be resolved. Membrane current changes are still filtered by the time-constant of the access resistance and the membrane capacitance. The investigator should not be misled by the rapid settling of the voltage step into thinking that this settling time represents the time resolution of the recording system.

Neither of these problems occur with the correction method of series-resistance compensation.

Usually, the type of series resistance compensation provided in a modern patch-clamp amplifier is a fixed combination of correction and prediction (9). In the recently developed Axopatch 200 patch clamp (Axon Instruments), the prediction and correction components of the circuitry can be independently set so that the researcher can optimize the technique that is most important for the experiment.

7. Patch-pipettes: practical considerations

7.1 Glasses

Glasses used for pulling patch micropipettes must be judged on at least three important criteria. These are:
- The current noise associated with the glass.
- The ease of pulling.
- Toxicity.

For minimal current noise, the glass should have a low loss factor. A high volume resistivity and a low dielectric constant may also result in low noise (19, 20). The so-called 'hard' glasses (aluminosilicate, e.g. Corning No. 1723, or its replacement, No. 1724) best satisfy these criteria for low noise. 'Medium' glasses (borosilicate, e.g. Corning No. 7052 and No. 7040) have slightly worse noise, except for Corning No. 7760, an exceptionally low loss borosilicate glass whose noise is comparable to the best hard glasses (James Rae and Richard Levis, personal communication). 'Soft' glasses (soda lime, e.g. Kimble No. R6) have the worst noise.

The medium and soft glasses are generally easier to pull. The high temperatures required to fabricate pipettes from hard glasses cause rapid degradation of the micropipette puller filaments. Furthermore, micropipettes pulled from hard glasses tend to be longer with a more gentle taper. With soft glasses, it is possible to pull bullet-shaped micropipettes. Since these are shorter with a steeper taper, the bulk resistance is lower and these micropipettes are more suitable for whole-cell recordings. These differences in ease of pulling are most acute when conventional micropipette pullers are used. Recent advances in microcprocessor-based electrode pullers allow electrodes to be fabricated in several heating and pulling cycles. These cycles can be tailored so that acceptable whole-cell electrodes can be pulled from almost any glass.

One exception to this grouping is the Corning No. 8161 glass. This pulls at a low temperature, seals well, and has noise as low or better than Corning No. 1723 aluminosilicate glass (19). However, this glass has recently been found in several cells to leach some compound which blocks several ionic currents (21, 22). This serves as a warning that it is not safe to assume that the micropipette glass is inert. Recordings should be made and compared using more than one glass type.

On the principle that the lowest noise is associated with the lowest loss materials, there is some speculation that quartz would be an excellent material for manufacturing patch micropipettes. Unfortunately, quartz is rarely used for this application because of the difficulty of fabrication. Quartz has a very high melting point, too high to be fabricated using a standard platinum heater filament in air. It is possible, however, that over the next few years pullers will become readily available to pull quartz micropipettes (see next section) and it will be interesting to see if there is any further reduction in noise to be gained. Sapphire has also been suggested as an alternative to quartz. It has the advantage of having a lower melting point than quartz but it suffers from the significant disadvantage of not being readily available as capillary tubing.

7.2 Pullers

Puller technology has advanced rapidly in the last few years. Some of the major enhancements of recent times are:

(a) *Air cooling of the filament.* First introduced by Brown and Flaming (23), this technique uses a timed blast of air or nitrogen to rapidly cool the filament. This enables more precise control of the tip geometry than is achieved when the filament is allowed to cool passively in room air with its variable drafts and humidity.

In 1987, Sutter Instrument Company (Novata, California, USA) improved the technique by replacing the compressed nitrogen supply with a low-pressure air delivery system. This system uses its own compressor and a solid-state pressure regulator. The sensitive solid-state regulator enables gas pressures to be controlled within 70 pascals. In contrast, a mechanical regulator on a nitrogen supply varies by up to 15 kilopascals between pulls. The improved pressure regulation contributes to improved consistency.

(b) *Velocity transducers.* Pullers typically change from a weak pull to a strong pull after the glass begins to melt. Traditionally, the transition from weak pull to strong pull has been triggered by the distance travelled as the two ends of the glass begin to pull apart. The distance travelled depends on the temperature of the glass as well as the rate at which the glass is heated.

The newest technique is to base the trigger point on the velocity with which the two ends of the glass pull apart after the glass softens (Suter Instrument Co., Novato, California, USA). For a given glass profile, this is dependent on the temperature of the glass, irrespective of the rate of heating. Thus, the pipette tip geometry is more predictable and repeatable. The benefit of this approach is more significant for patch micropipettes than for intracellular micropipettes.

(c) *Constant current power supplies.* The current to the filament and to the pulling solenoids has traditionally been supplied by transformers driven from the mains voltage. Since the mains voltage constantly varies, this is a source of significant inconsistency between pulls. Consistency is markedly improved by using constant-current supplies for the filament and the solenoids (24).

(d) *Microprocessor control.* Many pullers made in the last five years or so have incorporated microprocessor controllers. These enable complex pulling cycles to be programmed by the user. Certain combinations of pull strengths, of filament temperatures and of trigger points can be programmed to occur sequentially so that an optimum profile is created. This looping pull cycle technique is of particular benefit for fabricating patch pipettes.

An ancillary advantage of a programmable puller is that multiple programs can be stored. This makes it easier to share the puller among several researchers, although since most researchers regard micropipette fabrication as black magic, few are willing to share their puller. In this case, the multi-program capability can be used to store programs for different types of glass or different tip profiles.

The microprocessor controller is usually built into the puller. An alternative is to control the puller using an external computer (Mecanex, Nyon, Switzerland).

(e) *Quartz*. The most recent technology to emerge in micropipette pullers allows micropipettes to be pulled from quartz capillaries. Two techniques are being used. The first (25) uses a graphite filament to achieve the high temperatures required. Ordinarily, the graphite filament would oxidize in the air, so a continuous flow of argon is directed over the filament whenever it is heated.

The second system uses an infra-red CO_2 laser to heat the glass (Dale Flaming, personal communication). The output of the CO_2 laser is at 10.6 µm, very close to the resonant frequency of the SiO_2 bond (10.7 µm). Thus, absorption is extremely efficient. This technique has a number of advantages. First, no argon flow is required since there is no filament to oxidize. Second, the laser beam can be directed by computer-controlled mirrors to rapidly scan back and forth over the glass. The scanning pattern can be established by the microprocessor for an optimal heating pattern that will pull pipettes with the desired profile.

7.3 Holders

The main progress in holders in the last few years has been the recognition of the contribution of various design features to the current noise.

(a) It is most important to use the right material. This can only be determined empirically. The best material found to date is polycarbonate (J. Rae and R. Levis, personal communication; and ref. 26). This has even lower noise than Teflon. Other materials such as sapphire or low porosity ceramics might eventually prove to be better.

(b) The diameter of the pin used for the electrical connection should be small (e.g. 1 mm) and there should be a generous amount of insulation between it and the case (e.g. 4 mm).

(c) The holder should not be surrounded by a driven metal shield since this increases the current noise by virtue of the capacitance that it adds to the headstage input. (Note that from the noise point of view, the fact that the metal shield is driven does not help.)

(d) The holder should be designed so that the silver chlorided wire or the silver chloride pellet reaches down to near the tapering region of the micropipette so that the entire pipette does not need to be filled with solution. Filling the entire pipette increases the capacitance and may result in fluid leaking into the holder which causes further excess noise.

Another area of progress is the design of holders for perfusion of the filling solution. These holders have a thin perfusion pipette that is inserted into the patch micropipette. These systems suffer from increased noise compared with a well designed non-perfusion holder because there is a much larger surface area of electrolyte, in the micropipette itself, in the perfusion tubes and in the fluid reservoirs. A holder designed to minimize the stray input capacitance was described by Lapointe and Szabo (27).

8. Conclusion

The pace of technological progress in the instrumentation used for recording from single channels and from small cells has been rapid and will probably continue at a fast clip. Much progress in low-noise and wide-bandwidth recording has already taken place with the introduction of the original resistor-feedback technology followed by the low-noise capacitor feedback technology. In principle, it remains possible to lower the noise of the electronics by a substantial amount. Other areas where significant progress is expected to take place include increased computerization of the instruments, more powerful analysis software, and better glasses and pullers.

Acknowledgements

Thanks and appreciation are owed to the following persons: Richard Lobdill, for engineering discussions and for generation of some of the figures; Richard Levis and James Rae for technical input and for teaching the author about glasses, holders and capacitor-feedback headstages; Dale Flaming for sharing information about micropipette pullers; JoAnna Nguyen for assistance with the figures.

References

1. Neher, E. and Sakmann, B. (1976). *Nature*, **260**, 779.
2. Strickholm, A. (1961). *Journal of General Physiology*, **44**, 1073.
3. Strickholm, A. (1962). *Journal of Cellular and Comparative Physiology*, **60**, 149.
4. Neher, E. and Lux, H. D. (1969). *Pflügers Archiv für die Gesamte Physiologie*, **311**, 272.
5. Hamill, O. P., Marty, A., Neher, E., Sakmann, B., and Sigworth, F. J. (1981). *Pflügers Archiv für die Gesamte Physiologie*, **391**, 85.

6. Sakmann, B. and Neher, E. (ed.) (1983). *Single Channel Recording*. Plenum Press, New York.
7. Sigworth, F. J. (1986). *Federation Proceedings*, **45**, 2673.
8. Rae, J. L. and Levis, R. A. (1984). *Molecular Physiology*, **6**, 115.
9. Sigworth, F. J. (1983). In *Single Channel Recording* (ed. B. Sakmann, and E. Neher), p. 3. Plenum Press, New York.
10. Offner, F. F. and Clark, B. (1985). *Biophysical Journal* **47**, 142a.
11. Neher and Marty, (1982). *Proceedings of the National Academy of Sciences of the USA*, **79**, 6712.
12. Fernandez, J. M., Neher, E., and Gomperts, B. D. (1984) *Nature*, **312**, 453.
13. Lindau, M. and Neher, E. (1988). *Pflügers Archiv für die Gesamte Physiologie*, **411**, 137.
14. Joshi, C. and Fernandez, J. M. (1988) *Biophysical Journal*, **53**, 885.
15. Hodgkin, A. L., Huxley, A. F., and Katz, B. (1952). *Journal of Physiology*, **116**, 424.
16. Finkel, A. S. and Gage, P. W. (1985). In *Voltage and Patch Clamping with Microelectrodes* (ed. T. G. Smith, H. Lecar, S. J. Redman, and P. W. Gage), p. 47. Williams & Wilkins, Baltimore.
17. Giles, W. R. and Van Ginneken, A. C. G. (1985). *Journal of Physiology*, **368**, 243.
18. Armstrong, C. M. and Chow, R. H. (1987). *Biophysical Journal* **52**, 1333.
19. Rae, J. L. and Levis, R. A. (1984). *Biophysical Journal*, **45**, 144.
20. Rae, J. L., Levis, R. A., and Eisenberg, R. S. (1988) In *Ion Channels* (ed. T. Narahashi), p. 283. Plenum Publishing, New York.
21. Cota, G. and Armstrong, C. M. (1988). *Biophysical Journal*, **53**, 107.
22. Furman, R. E. and Tanaka, J. C. (1988). *Biophysical Journal*, **53**, 287.
23. Brown, K. T. and Flaming, D. G. (1977). *Neuroscience Journal*, **2**, 813.
24. Corey, D. P. and Stevens, C. F. (1983). In *Single-Channel Recording* (ed. B. Sakmann, and E. Neher), p. 53. Plenum Press, New York.
25. Munoz, J.-L. and Coles, J. A. (1987), *Journal of Neuroscience Methods*, **22**, 57.
26. Axopatch-1 manual (1985). Axon Instruments, Inc., Foster City, California, USA.
27. Lapointe, J.-Y. and Szabo, G. (1987). *Pflügers Archiv für die Gesamte Physiologie*, **410**, 212.

2

Distinguishing between multiple calcium channel types

MARTHA C. NOWYCKY

1. Introduction

The use of electrophysiological techniques to distinguish unique pathways for ion flow across biological membranes dates back to the pioneering work of Hodgkin, Huxley, and Katz. They successfully defined the changes in conductance which underlie action potentials and postulated the existence of independent, discrete channels for rapid Na^+ influx and a slightly delayed K^+ efflux. The strategy for discriminating between different ion channels electrophysiologically is based on finding differences in:

- kinetics, that is, the time- and voltage-dependence of channel opening and closing properties;
- ionic permeability and selectivity;
- pharmacological profiles.

A Ca^{2+} channel is defined as one that is strongly selective for Ca^{2+} ions, over the more numerous Na^+ and K^+ ions found in extracellular fluids. Since the advent of patch clamping techniques in the 1980s, it has become clear that there are multiple types of Ca^{2+} channels which are functionally and structurally related, but which differ in detailed aspects of their behaviour and properties. Eventually, subtypes of Ca^{2+} channels will be defined by differences in protein sequence and associated post-translational modifications; however, at this time such structural information is rudimentary. Therefore, Ca^{2+} channels continue to be segregated into types based on the differences mentioned above. However, they pose several special challenges to the investigator:

(a) There appears to be a spectrum of Ca^{2+} channel types. Ca^{2+} channel types are defined by a cluster of characteristics; many cells have channels which match most of the characteristics within one cluster, but differ in one or two properties.

(b) The voltage-ranges and time-courses of channel openings and closings for different subtypes often overlap, making separation based on any one kinetic criterion difficult. Additionally, the exact voltage-ranges and time-dependencies vary between cell types and across species.

(c) Most Ca^{2+} channels are subject to a great degree of biochemical modulation by intracellular mechanisms such as phosphorylation and thus one channel type may exhibit different kinetic behaviours under different metabolic conditions.

(d) No truly specific or invariably characteristic pharmacological tool is currently available for any calcium channel subtype.

The lack of clean diagnostic tests for Ca^{2+} channel types and their variable properties means that a spectrum of characteristics must be determined for all Ca^{2+} channels in a new preparation.

2. Major distinguishing features of currently identified calcium channel types

Most vertebrate neurons contain at least two types of Ca^{2+} channels, while some may have three, or even more. To illustrate the major distinguishing characteristics of Ca^{2+} channels, I will describe the three channel types in chick dorsal root ganglion neurons (see *Table 1* and refs 1–3). (Invertebrate neurons have Ca^{2+} channels which are difficult to categorize using vertebrate criteria and will not be considered in this review.)

2.1 Kinetic properties (voltage- and time-dependence of channel openings and closings)

2.1.1 Activation range

Calcium channels have been divided into high-threshold and low-threshold channel types, where 'threshold' refers to the voltage range over which channels open (are 'activated'). The two categories are defined relative to each other, and the exact range for each depends on details of external and internal solutions, as well as the cell type. Generally, ~ -20 mV can be considered an average threshold for high-threshold channels, while low-threshold channels may open as negative as ~ -70 mV. In *Table 1*, T is a typical, low-threshold channel, while both N and L are classified as high-threshold. Differences in the time to peak are usually not diagnostic of Ca^{2+} channel types.

2.1.2 Decay-time or inactivation rate

Rapidly-inactivating or 'transient' currents are those whose time-constant of decay during a maintained depolarization is between ~ 20 to ~ 100 msec (one time-constant or 'τ, tau' is the time it takes to reach $1/e$ or $\sim 37\%$ of the initial value). Slowly decaying or 'non-inactivating' currents have decay rates measured in the hundreds of milliseconds to seconds. Almost all low-

Table 1. Distinguishing features of the three types of vertebrate calcium channels in chick DRG neurons (from refs 1–3 and 14)

	T 'low-threshold'	N 'high-threshold'	L 'high-threshold'
Activation range (for 10 mM Ca)	Positive to −70 mV	Positive to −20 mV	Positive to −10 mV
Inactivation rate (0 mV, 10 mM Ca)	rapid (τ ~20–50 msec)	moderate (τ ~50–80 msec)	very slow (τ > 500 msec)
Inactivation range (for 10 mM Ca)	−100 to −60 mV	−120 to −30 mV	−60 to −10 mV
Deactivation rates (−80 mV, 10 mM Ca)	slow (τ = 1.5 msec)	(?)	fast (τ = 160 μsec)
Single-channel kinetics	late opening, brief burst, inactivation	long burst, inactivation	hardly any inactivation
Single channel conductance (110 mM Ba)	8 pS	13 pS	25 pS
Relative conductance	Ba = Ca	Ba > Ca	Ba > Ca
Dihdryopyridine sensitivity	no	no	yes
ω-CgTx block	weak, reversible	persistent	persistent(?)
Inorganic blocker sensitivity	Ni > Cd	Cd > Ni	Cd > Ni

threshold currents found thus far are rapidly inactivating (T for 'transient', Table 1). High-threshold currents either inactivate very slowly ('L' for 'long-lasting') or relatively rapidly N, for 'neither' or 'iNtermediate'). N-type channels exhibit a broad range of inactivation rates in different cells, which can overlap with either T- or L-type values at either extreme. If a channel is subject to Ca^{2+}-dependent inactivation, the use of Ba^{2+} as the charge carrier will produce dramatically slower decay values, since Ba^{2+} generally cannot substitute for Ca^{2+} in this process.

2.1.3 Steady-state inactivation range

Steady-state inactivation refers to the fraction of channels which are unavailable for opening at a given maintained or 'holding' potential. Generally, the voltage-range of inactivation is negative for low-threshold channels, (e.g. T is completely inactivated at −60 mV, and fully available at about −100 mV, Table 1) and positive for high-threshold, non-activating channels (e.g. L-type channels are completely inactivated at about −10 mV and fully available at −60 mV). High-threshold, inactivating channels, have ranges which are broader, and extremely variable between cell types.

2.1.4 Deactivation rates (tails)

The process of 'deactivation' refers to the rates at which Ca^{2+} channels close when the voltage is made negative after a depolarizing step opens the channels. Because of the increased driving force for Ca^{2+} entry at negative potentials, any current flowing through open channels is large and the closing of channels can be detected as a diminishing 'tail' current at the end of a test depolarization. Different Ca^{2+} channels close at different rates. Low-threshold channels (T-type) close more slowly and the value of τ is generally a few milliseconds. High-threshold, non-activating (L-type) channels have faster tails with τ values of several hundred microseconds. The rate of closure of N-type channels is controversial and probably overlaps with one of the other types. The lack of a clearly defined N-evoked tail current in dorsal root ganglion cells has been used as an argument against the existence of N-type channels.

2.2 Permeability

2.2.1 Throughput rate (chord conductance)

All known voltage-gated Ca^{2+} channels have extremely high throughput rates, passing as many as a million Ca^{2+} ions/second despite the presence of 10- to 50-fold greater concentrations of monovalent cations in physiological solutions. Within a negative voltage range (usually negative to 0 mV), the amplitude of single channel currents varies linearly with voltage and thus the slope or chord conductances of different channels can be easily compared. The conductance of the T-type channel is 8–10 pS in isotonic barium, that of the N-channel is 13–15 and that of the L-type has been reported as 20 to 28 pS (*Table 1*). These are the dominant conductance states of the channels; however, subconductance states have been reported for both the L- and N-type channels. The separation of conductances is not as clear if Ca^{2+} is used as the charge carrier (see Section 2.2.2).

2.2.2 Selectivity

All voltage-gated Ca^{2+} channels strongly select Ca^{2+} over monovalent cations, but have different selectivity for divalent cations. Low-threshold, T-type channels pass Ca^{2+} and Ba^{2+} equally well, or show a slight preference for Ca. In contrast, the high-threshold L- and N-type channels pass Ba^{2+} about twice as well as Ca^{2+}. The lesser permeability of T-type channels to Ba^{2+} is responsible for the better resolution of differences in single-channel amplitudes with this cation.

2.3 Pharmacology

2.3.1 Organic compounds

(a) Dihydropyridines (DHP): This class of compounds binds to high-threshold, non-inactivating channels (L-type). Most DHPs decrease Ca^{2+} channel activity (e.g. nifedipine, nitrendipine, nisoldipine) and thus are referred to as antagonists, while others increase channel activity (BayK 8644, CGP 28392). The DHPs present several difficulties when used as a dissecting tool. First, L-type Ca^{2+} channels in various cell types have different sensitivites to individual DHPs. Second, neuronal channels are relatively insensitive compared to skeletal or cardiac muscle and require higher concentrations which may have additional non-specific effects. Third, the binding, and thus the efficacy, of these compounds has complex voltage- and use-dependence. Fourth, in some tissues, T-channels are partially blocked by some of the DHPs (felodipine, ref. 4). Nevertheless, when used cautiously, DHPs are useful in identifying (or ruling out) L-type channels.

(b) Verapamil, a phenylalkylamine, selectively blocks high-threshold channels in rat and chick DRG cells (5). It is not clear if it discriminates between L-type and N-type channels. Further work in a range of neuronal types is necessary to determine its selectivity.

(c) ω-conotoxin (ωCgTx): This peptide toxin irreversibly blocks high-threshold channels in vertebrate neurons. Several recent reports suggest that the toxin may be completely specific to N-type Ca^{2+} channels; however, this must be substantiated in a spectrum of species and cell types.

(d) Three organic compounds have been reported to selectively block low-threshold, T-type channels. These are amiloride (6), a potassium-sparing diuretic, tetramethrin (7), an insecticide, and octanol (8). Results with amiloride and tetramethrin have been presented only for neuroblastoma cells, while octanol has been studied in inferior olivary neurons.

2.3.2 Inorganic cations

Except for Ba^{2+} and Sr^{2+} which permeate voltage-gated Ca^{2+} channels, most other divalents and trivalents of the transition element series block Ca^{2+} permeation through Ca^{2+} channels. Several of these block different calcium channels types with differing efficacies. The most thoroughly studied inorganic blockers are Cd^{2+}, which preferentially blocks high-threshold channels, and Ni^{2+}, which preferentially blocks low-threshold channels (*Table 1*). It is very important to note that the individual K_ds of these and other inorganic blockers for the different channels vary by only about one log unit (9). At higher concentrations, all of them block all types of Ca^{2+} channels, as well as certain other voltage-gated channels.

3. Whole-cell currents

3.1 General strategy

- Select solutions which isolate Ca^{2+} currents and block contaminating currents.
- Apply appropriate voltage-step protocols to obtain a maximum of kinetic information.

- When possible, apply some pharmacological tool or change the permeant cations.

3.2 Extracellular solutions

(See *Table 2* for sample solutions.)

3.2.1 Extracellular cations

(a) Divalent charge carriers; Ca^{2+}, Ba^{2+}, and Sr^{2+} are the three divalent cations most commonly used to study ion flow through Ca^{2+} channels. The physiological concentration of calcium in the extracellular fluid of vertebrates is 2 mM. However, in many cells Ca^{2+} currents are small, and the experimental concentration is raised to improve the signal-to-noise ratio. The amplitude of calcium currents increases almost linearly with concentrations between 2 to 10 mM, and approaches saturation above 10 mM. Ba^{2+} provides two advantages: it blocks many potassium currents and it permeates almost twice as well through high-threshold channels. On the other hand, the inactivation kinetics of currents may be slowed by the use of Ba^{2+} because this ion does not support Ca^{2+}-dependent

---

Table 2. Sample solutions (all numbers are in millimoles)

Whole-cell current recordings

Embryonic chick DRG (ref. 2)
Extracellular before seal:	130 NaCl, 5 CaCl$_2$, 5 KCl, 1 MgCl$_2$, 10 glucose, 10 Hepes, pH 7.3 with NaOH
Extracellular after seal:	153 TEA-Cl, 3–10 CaCl$_2$ or BaCl$_2$, 10 glucose, 10 Hepes, pH 7.3 with TEA-OH and 200 nM TTX
Intracellular:	100 CsCl, 10 Cs-EGTA, 5 MgCl$_2$, 40 Hepes, 2 ATP, 0.25 cAMP, pH 7.3 with CsOH

Embryonic rat DRG (ref. 10)
Extracellular:	118 Tris Cl, 14.8 CaCl$_2$ or BaCl$_2$ or SrCl$_2$, 2 MgCl$_2$, 5 Tris Cl, pH 7.4.
Intracellular:	150 Tris phosphate, pH 7.4.

Adult bullfrog DRG (ref. 11)
Extracellular:	3 BaCl$_2$, 154 TEACl, 10 Hepes, pH 7.4 with TEA-OH
Intracellular:	126 Cs glutamate, 9 EGTA, 4.5 MgCl$_2$, 9 Hepes, 3.6 MgATP, 14 creatine phosphate (tris salt), 1 GTP (Na$^+$ salt), 50 U ml^{-1} creatine phosphokinase, pH 7.4 with CsOH

Single-channel recordings

Embryonic chick DRG (ref. 3)
Bath:	140 KAspartate, 10 K-EGTA, 1 MgCl$_2$, 10 Hepes, pH 7.4 with KOH
Pipette:	110 BaCl2, 10 Hepes, pH 7.4 with Ba(OH)$_2$

inactivation. It is generally best to characterize the currents with both Ba^{2+} and Ca^{2+}.

(b) Na^+: To eliminate current flow through Na^+ channels, either extracellular Na^+ should be replaced with a large, impermeant cation or tetrodotoxin (TTX) should be added in high enough concentrations to block all Na^+ channels. Commonly used substitutes are choline, N-methyl-D-glucamine (NmDg), tetraethylammonium, or Tris. Neuronal Na^+ channels are blocked by nM concentrations of TTX, however, the toxin can be used in excess (e.g. 1 μM).

3.2.2 Extracellular anions

Chloride is the most commonly used anion in extracellular solutions, since it is the physiological anion. However, at times it is replaced with larger organic anions such as aspartate, glutamate, or phosphate.

3.2.3 Other components

(a) K^+ channel blockers: Most neurons have a large variety and number of voltage-gated and Ca^{2+}-gated potassium channels. These currents flow outward over the same voltage-range at which calcium channels are activated and obscure the smaller Ca^{2+} currents. Internal K^+ is usually replaced with a larger, impermeant cation (see Section 3.3.1.). Additionally, it is often necessary to include one or more of the following K-channel blockers externally: Ba^{2+} (as the divalent cation charge carrier); tetraethylammonium (TEA); 4-aminopyridine (4-AP). The exact concentration must be determined empirically for a given neuron.

(b) Standard salt solutions will also contain a pH buffer (Hepes is effective over the physiological range), glucose, and $MgCl_2$.

3.3 Intracellular solutions

3.3.1 Intracellular cations

The choice of the major cation is usually based on its ability to block K^+ channels. Cs^+ is frequently used, since it does not permeate through many (but not all) K^+ channels. Additionally, some of the K^+ can be replaced with larger organic cations such as TEA or NmDg. However, these should be used cautiously, since Trautwein's group reported that NmDg modifies the kinetics and voltage-dependence of Ca^{2+} channels in heart cells (12).

3.3.2 Intracellular anions

Although Cl^- is frequently used experimentally as the major anion, most cells contain only a few millimoles of this anion. Aspartate, glutamate, and phosphate have been used to replace Cl^-. Fl^-, which is popular for stabilizing whole-cell recordings of Na^+ currents and many ligand-gated channels, 'kills' most types of Ca^{2+} channels.

3.3.3 Ca^{2+} buffers

It is critical to buffer intracellular Ca^{2+} ions to low levels, both because high $[Ca^{2+}]_i$ is deleterious to cells and because many Ca^{2+} channels are subject to Ca^{2+} mediated inactivation. Common buffers are EGTA and the recently developed BAPTA, which is more selective and more rapid than EGTA.

3.3.4 pH buffers

Intracellular pH is commonly controlled by Hepes, although phosphate buffers have also been used.

3.3.5 Maintaining channel functionality

In contrast to many other voltage-gated channels, Ca^{2+} currents gradually decrease during whole cell recordings, a phenomenon called 'rundown'. Rundown is probably caused by the gradual loss of channel phosphorylation and/or other biochemical processes as the cell contents are equilibrated with the pipette solution. The following compounds are useful in minimizing rundown. Mg·ATP and cAMP maintain the activity of the cAMP-dependent protein kinase. GTP activates G proteins. Leupeptin, an endogenous protease inhibitor, in some cells eliminates irreversible rundown resulting from channel degradation. High energy regenerating systems (e.g. creatine phosphokinase) can be helpful.

3.4 Pulse protocols

The basic building blocks for constructing voltage protocols which elicit kinetic differences between currents are:

- the steady-state or 'holding' potential
- test pulse potential
- test pulse duration
- test pulse interval
- tail potential

Appropriate pulse protocols constructed from these blocks provide information on parameters of activation, inactivation, recovery from inactivation, and deactivation.

3.4.1 Activation kinetics

To study the properties of activation, Ca^{2+} currents are elicited at a series of test potentials (TPs) from a constant holding potential (HP; *Figure 1*). Test pulses should be obtained at 5 or 10 mV increments, and should range from below the threshold to above the reversal potential (usually between ~-70 to $+100$ mV). To isolate the L-type current, HP is held relatively positive (~-30 to -60 mV). To study the rapidly decaying currents, the HP is made more negative (~-80 to -100 mV; determined from inactivation experi-

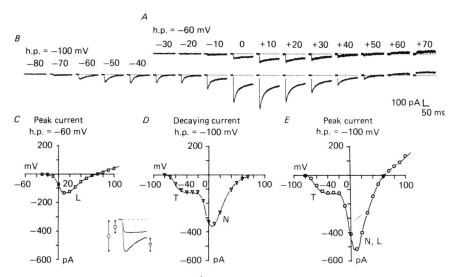

Figure 1. Three components of Ca^{2+} current observed under whole-cell recording conditions. A and B, currents evoked with pulses from HP = −60 mV (A) and −100 mV (B) to TP between −80 and +70 mV. C, peak current plotted against TP for current traces in A. D, magnitude of the decaying current (measured by as indicated by the triangle in the inset). E, peak current for traces in B. (From ref. 2.)

ments, Section 3.4.2) and some subtraction procedure must be used to eliminate L-type current. Since all the current types are seen from negative potentials, it's best to obtain the activation series first from a positive HP, since this is counter to the direction of rundown.

The duration of the test pulse is determined by the decay rates of transient and non-inactivating currents. It is often convenient to use one duration that is adequate for full decay of the transient current, then another which is about 5–10 times as long to determine the time-course of the long-lasting currents. Since full decay of the L-type current may require many seconds and maintained depolarizations are generally harmful to cells, it is often impractical to observe complete inactivation of long-lasting currents.

For most Ca^{2+} channels recovery from inactivation is extremely slow, so the interval between test pulses should be as long as is feasible given the general viability of the cells and stability of the seals and currents. In almost no case should test pulses be given more frequently than 1/sec. Many labs use interpulse intervals of 3 to 10 sec; however, even 1 min may not be enough to obtain full recovery.

In some cells, pre-pulses can be used to isolate the current (e.g. see ref. 13). In this case, the cell is held at a negative potential and transiently depolarized to a potential which elicits the low-threshold current. After full inactivation occurs, the TP is stepped more positive to elicit the high-threshold current.

This method can only be used if the following conditions are met: (a) The low-threshold current must be activated and inactivated at a TP which is not strong enough to activate the high-threshold current; and (b) The high-threshold current must not be significantly inactivated by the prepulse.

3.4.2 Steady-state inactivation

To determine the voltage-range of steady-state inactivation, the TP is kept constant while the HP is varied (*Figure 2*). For low-threshold channels, it is sometimes possible to use a TP which activates only that channel. Otherwise, some other method of isolating individual current types must be employed. The HP should be varied in 5 or 10 mV intervals from about −120 to 0 mV and the cell should be held for a minimum of 10–30 sec at each potential. If the preparation is robust, longer intervals can be used. In theory, it should be possible to obtain similar inactivation curves whether going from positive to negative potentials, vice versa, or moving randomly between potentials. In practice, many calcium channels undergo irreversible 'rundown' (perhaps degradation) at positive potentials. Therefore, it is frequently better to start at a negative HP and move in a positive direction, or make pseudo-random changes.

3.4.3 Deactivation (tails)

Since different Ca^{2+} channels close in a characteristic, first-order manner, tail currents can be fit with one or more exponentials, with each exponential reflecting the rate of closure of one type of channel. Tail current analysis can be extremely useful to address the kinetics of channel opening, closing, and inactivation (for an elegant example, see ref. 14). However, the technique requires cautious and conservative use. First, the recorded 'tail currents' can be contaminated with contributions from other channels, particularly Ca^{2+}-activated channels. Second, since most Ca^{2+} channels close quickly at negative potentials (< 1 to a few milliseconds), the speed of the voltage clamp, saturation of the recording electronics, and quality of spatial clamp must all be carefully controlled (15). Finally, the exponential rates of closing of two channel types should vary about three- to fivefold or more, to be adequately resolved.

3.5 Permeability

Replacement of the charge-carrying divalent cation in the external medium during an experiment will change the amplitude of a given current depending on the channel types' relative permeability to those divalents. Extreme care should be taken to fully exchange the bath, since many Ca^{2+} channels exhibit the so-called 'mole fraction effect' where mixtures of two permeant ions produce smaller currents than an equal concentration of either of the two ions (16).

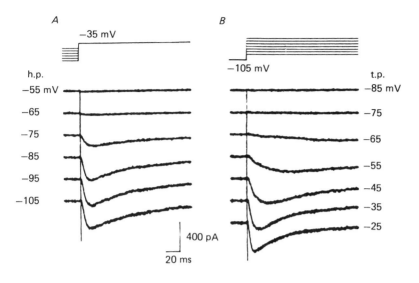

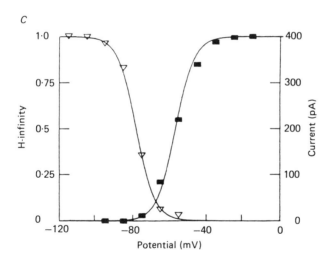

Figure 2. Voltage-dependence of inactivation and activation of T currents. *A*, isolated T currents evoked by depolarizations to TP = −35 mV from various HP. *B*, T currents elicited by depolarizing pulses from HP = −105 mV to different TP. *C*, triangles: peak current amplitudes, normalized to the maximal current obtained at HP = −115 mV, plotted against HP (data from *A*). Data were fitted with a smooth curve derived from the Boltzmann relation, $I/I_{max} = [1 + \exp((V-V_{1/2})/k)]^{-1}$ where $V_{1/2} = -78$ mV and $k = 5$ mV. Squares: peak current amplitudes, normalized to maximal current obtained at TP = −15 mV, plotted against the TP (from data in *B*). Data are fitted with a smooth curve $I/I_{max} = [1 + \exp((-V-V_{1/2})/k]^{-1}$ where $V_{1/2} = -51$ mV and $k = 6.5$ mV. (From ref. 2.)

3.6 Pharmacology

3.6.1 Dihydropyridines

A positive response to a DHP is almost always diagnostic of an L-type Ca^{2+} channel, although a few (DHPs), such as felodipine, may have effects on the T-type channel. The DHP class of compounds contains both blockers and promoters of L-type channel activity. Some of the DHPs are racemic mixtures, in which one enantiomer is a pure agonist, while the other is a pure antagonist. Whenever these are available they should obviously be used in their pure form.

Since most DHPs bind more strongly to the inactivated form of the channels, the cell potential should be relatively positive (< -50 mV). Block by antagonists is most easily seen during a test pulse which elicits a large Ca^{2+} current. In contrast, the effect of DHP agonists can be detected more easily by using TPs just at the foot of the activation curve since these compounds shift the L-type channel activation curve to the left. Alternatively, since most of the agonists greatly slow the channel deactivation rates, tail currents can be examined for a pronounced slowing and increased amplitude.

The lack of a response to DHPs is more problematic. The channel under investigation may be truly DHP-insensitive. However, DHP binding to L-type channels exhibits complex use- and voltage-dependence (17). Additionally, L-type channels from different cell types have markedly varied affinities for different DHPs. Thus, in the absence of a DHP response, a number of compounds under a variety of pulse-protocols must be tested.

The DHPs are all very lipid soluble and water insoluble. They are generally made up as stocks in the 1 to 50 mM range in ethanol or DMSO. These are then diluted to the final concentration range of nM to μM. Appropriate controls for the solvents must be performed. Some DHPs are light-sensitive and suitable precautions must be taken. A list of some DHPs which have effects on various L-type channels is provided in ref. 14.

3.6.2 ω-conotoxin

This 27 amino acid peptide irreversibly blocks high-threshold channels in neuronal preparations. It may be specific for N-type calcium channels; however, this must be determined in a variety of cells. The peptide may block L-type channels reversibly in endocrine cells. Any effects on low-threshold channels are small and transient.

ω-CgTx is freely water soluble and very stable, retaining full potency for days at room temperature. In powdered form it can be kept in the freezer for many months. It is applied in concentrations of nM to low μM. The toxin does not bind to channels if the divalent cation concentration is high (e.g. no block at 10 or 20 mM Ca^{2+} or Ba^{2+}), so low concentrations of divalent cations must be used.

3.6.3 Inorganic cations

Different Ca^{2+} channels can be distinguished on the basis of effectiveness of blockade by the inorganic cations, Cd, Ni, Gd, and Mg. These are simply added to the solution in appropriate concentrations (e.g. 100 µM cadmium blocks high-threshold, but not low-threshold channels, ref. 2). One complication is that because of their charge, the block produced by these ions displays voltage-dependence. Therefore it is necessary to compare block of low- and high-threshold channels at similar test potentials (19).

4. Whole-cell: data analysis

4.1 Strategy for isolation

If the activity of a single-channel type can be obtained by an experimental isolation procedure, such as inactivating all but one of the currents or recording in the presence of a selective blocker, the activity of the other current types in that cell may be obtained by subtraction. Alternatively, since the decay rate of the L-type current is very slow, if the other current types decay relatively rapidly, the peak current can be measured and compared to the current at the end of a long pulse (e.g. *Figure 1*).

4.2 Activation plots

The most common type of plot is the current–voltage relationship (*I–V*) where peak amplitude is plotted against the test depolarization (*Figure 1*). However, this data is a combination of both the activation properties and the conductance at any given potential. Another useful format is a plot of the fraction of channels activated at a given potential. This can be obtained by instantaneous tail measurement, which reflect current flow through all open channels at a negative potential before any have been able to close. The resulting curve can be fit with a Boltzmann distribution (*Figure 2C*). This empirical fit provides two values which make comparisons between different currents much simpler: the $V_{1/2}$ value which gives the potential at which one half of the channels are activated, and the k value which indicates the steepness of activation to the voltage-relationship.

4.3 Inactivation plots

Peak current amplitudes elicited at different HPs can be plotted directly or normalized to the maximum current to provide an 'H-infinity' plot (*Figure 2C*). Boltzmann distributions can be fit to the data points, as discussed above, allowing the determination of both the midpoint value, $V_{1/2}$, and k, the steepness factor.

4.4 Measuring rates of decay

The rate of decay of a current is usually expressed as τ, which is the time at which the current has fallen to $1/e$ or 37% of its original value. After appropriate isolation, the current trace is linearized by transforming the raw data into the corresponding natural log values, and is then fit with a least-squares or other procedure. The rate of decay often has a strong voltage-dependence and should be determined over a broad voltage range.

If two currents have very widely separated rates of decay (> fivefold) it is possible to obtain the rates of decay for both from single traces. Various mathematical routines such as the Levenberg–Marquardt can be used for fitting two exponentials to a single trace. Although it is mathematically possible to obtain three (or even more!) exponentials for a single trace, in practice there are rarely enough data points to justify this procedure. Instead, one, or at most two, current types should be isolated by some strategy and then fit.

5. Single-channel currents

5.1 General strategy

Single-channel recordings offer several advantages over whole-cell recordings.

(a) In the whole-cell mode, inadequate space clamp of cells that are large or are not spherical (e.g. the many branching neurites of neurons) can cause severe distortion of the apparent kinetics of the different current types. No such problems are encountered with single-channel recordings.

(b) Information on distribution and clustering of channels can be obtained.

(c) The slope conductance of a channel can be accurately determined from direct measurements of channel open size.

(d) Details of channel behaviour such as probability of opening, or open and closed times, can often only be obtained by direct analysis of single-channel records:

The disadvantages are:

(a) Since most patches contain only one or a small number of channels, tens or even hundreds of test pulses must be averaged to obtain information on the voltage- and time-dependent properties of the channels. This requires long, stable, and tedious experiments.

(b) Many of the pharmacological tests which can be used with whole-cell recordings are not feasible or are difficult. For example, it is often impossible to obtain control and wash recordings for compounds which act on the outside of the channel, unless the patch pipette can be perfused.

However, the correlation of single channel data with information obtained from whole-cell recordings is very powerful and both techniques should be used in combination whenever possible.

The general strategy is similar to that for whole-cell recordings.

(a) The pipette solution is selected to maximize the single-channel currents and to block flow through other types of ion channels. The bath solution can be a physiological saline or a solution chosen to clamp the cell resting potential.

(b) Test protocols are applied which determine various time- and voltage-dependent kinetic properties.

(c) Various compounds are tested to determine the pharmacological profiles of the channels.

Calcium channels usually must be recorded using the cell-attached variant of the patch-clamp method since most voltage-gated calcium channels rapidly 'rundown' if the patch is excised. An exception are the T-type channels of some cells, which may continue to function for as much as 15–30 min after excision.

5.2 Pipette solution

(See *Table 2* for example solutions.)

5.2.1 Pipette cations

The pipette solution is the 'extracellular' solution seen by the channel in the cell-attached mode. The divalent cation which is the charge carrier can be Ca^{2+}, Ba^{2+}, or Sr^{2+}. To resolve single channel openings, the concentrations of these must be much higher than the physiological levels of calcium. It is common to have the divalents completely replace monovalent cations in the solution. Ba^{2+} is popular since (a) it blocks many types of voltage-gated K^+ channels, (b) does not activate many of the Ca^{2+}-dependent K^+ and Cl^- channels in cell membranes, and (c) produces a larger amplitude current in those channels which pass Ba^{2+} better than Ca^{2+}.

5.2.2 Pipette anions

Chloride is the most common anion used; however, larger organic anion can be used if Cl^- channels are a problem (see Section 3.2.2).

5.2.3 Other components

K^+ channel blockers often need to be included in the pipette, even though it contains high Ba^{2+} levels (see Section 3.2.3). TTX should be included in the pipette to be confident that there is no flow through voltage-gated Na^+ channels, even in the absence of Na^+ in the pipette solution.

The only other solution component needed is a pH buffer, such as Hepes.

5.3 Bath solution

5.3.1 Physiological saline

This is the best choice for cells which have relatively stable resting potentials, do not exhibit spontaneous activity, and are large enough that individual channel openings do not depolarize the whole cell (e.g. cardiac myocytes). The resting potential of the cells can be determined either from direct whole cell current clamp recordings, or estimated from the amplitude of single channel openings, once the value of these is determined for a given test potential.

5.3.2 Non-physiological solutions

Many cells, particularly neurons, do not meet these criteria. Spontaneous activity can be eliminated by including TTX in the bath solutions. However, some neurons may have several stable 'resting' potentials or very high input resistances so that significant depolarization results from a few channel openings. In these cases, it is critical to somehow clamp the cell membrane potential. The ideal method is to use a different pipette which monitors or clamps the resting potential. This is not feasible for small or fragile cells. An alternative is to artificially 'zero' the cell by including high K^+ in the bath solution so that the cell membrane potential is known to be zero and stable. This method has the disadvantage that prolonged depolarization is deleterious to the cells in the bath and the preparation must be changed frequently.

5.4 Conductance (permeability)

One of the characteristic 'fingerprints' of different Ca^{2+} channel types is the different amplitudes of single-channel currents (*Figure 3*). If possible, this data should be obtained from patches containing only one type of channel. Since Ca^{2+} channel conductance is often not linear throughout the voltage range, a broad range of voltage values should be tested.

5.5 Pulse protocols

The procedures for determining the voltage- and time-dependent properties of single-channel activity are essentially identical to those used for whole-cell currents (*Figure 3*). However, since most patches contain only one or a few channels, large numbers of sweeps have to be averaged for any protocol, making some of the experiments impractical.

5.6 Pharmacology

5.6.1 Dihydropyridines

Since DHPs are very lipid soluble, they permeate easily through the membrane. Both agonists and antagonists can be added to the bath and will rapidly act on channels inside the pipette. The agonists can also be included

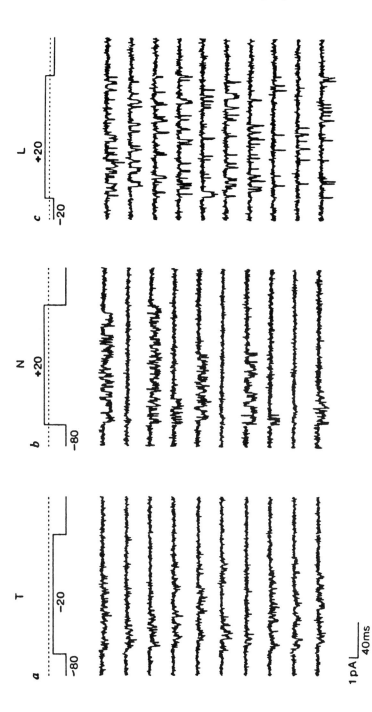

Figure 3. Three types of unitary Ca^{2+} channel activity seen in cell-attached patch recordings with Ba^{2+} as the charge carrier. *a*, T-type channel activity; *b*, N-type channel activity; *c*, L-type channel activity in three separate patches. HP and TP in for each patch indicated in voltage protocols above current traces. T-type and N-type activity disappeared if the HP was positive (e.g. −20 mV), while L-type activity persisted. Within each panel traces are consecutive. (From ref. 1.)

within the pipette: the L-type channel activity in their presence is so dramatically and characteristically increased (long openings, often outlasting the test potential), that it is easy to separate out L-type from other Ca^{2+} channels (*Figure 4A*).

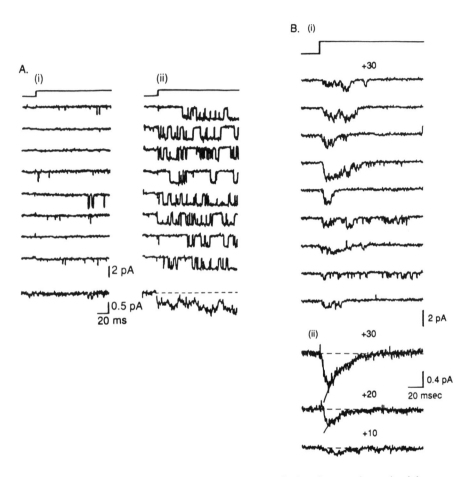

Figure 4. A. Terminal-attached recording from an isolated neuro-hypophysial nerve terminal. L-type channel activity before (i) and after (ii) addition of BayK 8644 to the bath. Sequential traces shown above and corresponding averaged current shown in lowest trace. Patch was depolarized to TP = ~−20 mV from HP = ~−80 mV. **B.** Inactivating, high-threshold channel from same preparation. Top nine traces are consecutive, representative sweeps elicited by steps from −80 to +30 mV. (ii) Three averaged currents from the same patch, stepping to the indicated potentials (averages of 124, 47, and 83 sweeps). Averages at +20 mV and +30 mV each were fitted with a single exponential with τ_{decay} = 16 and 22 msec, respectively. The relative probability of opening at +30 mV declined from 0.03 to 0.009 when HP was changed from −80 to −30 mV and increased to 0.038 after returning to −80 mV). (From ref. 22.)

5.6.2 ω-conotoxin

Since ω-CgTx does not bind to the Ca^{2+} channels in the presence of high concentrations of divalents, and since it does not permeate the membrane but must act directly on the outside of the channel, single channel experiments are quite difficult. However, in their original report, McClesky et al., were able to use the outside-out patch conformation with very large 'bag-like' patches (20).

5.6.3 Other

Inorganic blockers are not useful for distinguishing Ca^{2+} channel types at the single channel level.

6. Single-channel: data analysis

6.1 Kinetic analysis

Most of the strategies used to isolate and characterize whole cell currents apply to single channels as well. If one channel type is present in the patch at high density, a macroscopic current is produced which can be analysed exactly like the corresponding whole cell current. More frequently, the number of channels in a patch is very low and many test pulses have to be averaged to obtain records with discernible kinetic features (*Figure 4B*). Because of the relatively low signal to noise ratio in most Ca^{2+} channel recordings small fluctuations in the baseline can overwhelm the actual data. To avoid this, a baseline should be determined for each trace and set to zero before averaging. For software strategies to determine baselines, see ref. 21.

6.2 Conductance

The amplitude of single channel openings can be determined 'by eye,' but provided there are enough openings, amplitude histograms should be generated. These can be fit with the Gaussian function to give more accurate values. To calculate the slope conductance single-channel amplitudes are plotted against voltage. Some problems that will be encountered are:

- Presence of more than one type of channel in the patch.
- Presence of subconductance states for some types of Ca^{2+} channels.
- Brief openings or 'flicker' which give artefactually low amplitude values. This can be minimized for L-type channels by using the DHP agonists which greatly prolong channel openings.

7. Summary

While in different cell types and species the L- and the T-type currents appear to be reasonably similar, though not completely identical, high-threshold, inactivating channels (N-type) vary considerably and probably constitute a family of closely related proteins. This raises the question of nomenclature: how different should a channel be to deserve a separate designation?

To some extent the answer is a matter of aesthetics: one tries to strike a balance between keeping the number of channel types to a useful minimum, while at the same time not obscuring the existence of important differences. Although it is sometimes discouragingly tedious, whenever a new preparations is encountered a full range of the properties of calcium channels should be determined.

8. Supplies

The majority of the compounds and salts mentioned in this chapter are easily available from major chemical suppliers such as Sigma or Aldrich.

Many of the dihydropyridines can only be obtained from their pharmaceutical manufacturers as a gift. Two exceptions are:
    BayK 8644 – Calbiochem
    Nifedipine – Calbiochem or Sigma
Others:
    ω-conotoxin – Peninsula Laboratories
    tetrodotoxin – Calbiochem
    leupeptin – Calbiochem
    BAPTA – Calbiochem
    CsOH – Aldrich

References

1. Nowycky, M. C., Fox, A. P., and Tsien, R. W. (1985). *Nature*, **316**, 440.
2. Fox, A. P., Nowycky, M. C., and Tsien, R. W. (1987). *Journal of Physiology*, **394**, 149.
3. Fox, A. P., Nowycky, M. C., and Tsien, R. W. (1987). *Journal of Physiology*, **394**, 173.
4. Van Skiver, D. M., Spires, S., and Cohen, C. J. (1989). *Biophysical Journal*, **55**, 593a.
5. Boll, W. and Lux, H. D. (1986). *Experimental Brain Research*, **14**, 104.
6. Tang, C.-M., Presser, F., and Morad, M. (1988). *Science*, **240**, 213–215.
7. Tsunoo, A., Yoshi, M., and Narahashi, T. (1985) *Biophysical Journal*, **47**, 433a.
8. Llinas, R. and Yarom, Y. (1986). *Neuroscience Abstracts*, **12**, 174.
9. Narahashi, T., Tsunoo, A., and Yoshii, M. (1987). *Journal of Physiology*, **383**, 231.
10. Fedulova, S. A., Kostyuk, P. G., and Veselovsky, N. S. (1985). *Journal of Physiology*, **359**, 431.
11. Bean, B. P. (1989). *Nature*, **340**, 153.
12. Malecot, C. O., Feindt, P., and Trautwein, W. (1988) *Pflügers Archiv für die Gesamte Physiologie*, **411**, 235–242.

13. Vivadou, M. B., Clapp, L. H., Walsh, J. V., and Singer, J. J. (1988). *Federation of Societies for Experimental Biology Journal*, **22**, 2497.
14. Swandulla, D. and Armstrong, C. M. (1988). *Journal of General Physiology*, **92**, 197.
15. Armstrong, C. M. and Chow, R. W. (1987). *Biophysical Journal*, **52**, 133.
16. Hess, P. and Tsien, R. W. (1984). *Nature*, **309**, 453.
17. Sanguinetti, M. C. and Kass, R. S. (1984). *Circulation Research*, **55**, 336.
18. Scriabine, A. (1987). In *Structure and Physiology of the Slow Inward Calcium Current* (ed. J. C. Venter and D. Triggle), p. 51. Alan R. Liss, New York.
19. Lansman, J. B., Hess, P., and Tsien, R. W. (1986). *Journal of General Physiology*, **88**, 321.
20. McCleskey, E. W., Fox, A. P., Feldman, D. H., Cruz, L. J., Olivera, B. M., Tsien, R. W., and Yoshikami, D. (1987). *Proceedings of the National Academy of Sciences of the USA*, **84**, 4327.
21. Sachs, F. (1983). In *Single-Channel Recording* (ed. B. Sakmann and E. Neher), p. 265, Plenum Press, New York.
22. Lemos, J. R. and Nowycky, M. C. (1989). *Neuron*, **2**. 1419.

3

Physiological approaches to the study of glutamate receptors

C. E. JAHR and G. L. WESTBROOK

1. Introduction

Our understanding of glutamate receptors and their role in central excitatory neurotransmission has dramatically increased in the past five years. This has been driven by rapid technical improvements in the pharmacological agents available for studying glutamate receptors, as well as by the refinement of *in vitro* preparations. The purpose of this chapter is to survey some of the methods that have been useful in studies of glutamate receptors and ion channels in the past few years. Obviously, many of these techniques are common to the study of any receptor or ion channel, especially in the central nervous system (CNS). Thus no attempt will be made here to comprehensively review all the techniques that may have application to glutamate receptor studies. Rather we will focus on techniques which have influenced major issues in excitatory amino acid physiology and pharmacology.

Some of the experimental questions which have received most interest include which post-synaptic glutamate receptors are involved in excitatory neurotransmission. In addition, the roles of modulators of *N*-methyl-D-aspartate (NMDA) channel such as glycine and magnesium have been studied extensively, as have drugs which act as open channel blockers of the NMDA channel. The flux of calcium ions through the NMDA channel and its significance for cellular function, particularly as it relates to long-term potentiation (LTP) in the hippocampus has also been examined. Finally, there is increasing interest and progress in the study of excitatory amino-acid receptor activation of second-messenger pathways, presumably through G protein mechanisms. In the last section techniques for studying structure and function of cloned glutamate receptors using expression systems such as the *Xenopus* oocytes are briefly described. These studies are just beginning, but are likely to be major experimental areas for the future.

2. Experimental preparations

In vertebrates, glutamate receptors are restricted primarily to the CNS. Most CNS neurons express at least one, and generally several, glutamate receptor subtypes. As a result, preparations from all regions of the central nervous system have been used in studies of glutamate receptor function. In general, it appears that receptor subtypes at all levels of the neuroaxis have similar properties. However, it is likely that more detailed examination will reveal both regional and developmental differences of functional importance. Although *in vivo* preparations were used initially to study glutamate receptors, the lack of physical access to individual cells and the importance of voltage-clamp techniques make *in vitro* preparations preferable for studies of glutamate receptor conductance mechanisms. Three types of *in vitro* preparations have been used most frequently for physiological studies of glutamate receptors: tissue slices, acutely dissociated neurons and primary neuronal cell cultures. Each preparation has particular experimental advantages. For example, the slice preparation is particularly useful for studies of synaptic mechanisms as intact pre- and post-synaptic elements are present (see ref 1, for example). The major advantage of either acutely dissociated cells or primary cultures is the ability to record from individual cells and apply known concentrations of drugs to the cell membrane. The slice culture technique introduced by Gähwiler has some of the advantages of dissociated cells, but has seen limited use in studies of glutamate receptors. In addition, acutely dissociated cells or cultured cells can be voltage-clamped using patch-clamp approaches which greatly facilitate studies of ion channel and receptor mechanisms. This latter advantage of isolated cells (or of the slice culture) has diminished somewhat with the development of patch-clamp techniques for acute thin-slice preparations (2, 3).

Primary cultures of CNS neurons are virtually ideal preparations for studies of glutamate receptor ion channels. Methods for culturing neurons from various CNS regions such as hippocampus, spinal cord, or cerebellum are now well-described and relatively easy to perform. Most cultured neurons are excited by application of L-glutamate, and most have multiple receptor subtypes (4). Thus detailed pharmacological and physiological analyses are quite feasible. The general problems with use of cultured neurons include identification of cell type and the possible effects of culture conditions. Cell identity in culture can be determined by using *in vivo* retrograde labelling techniques to identity projection neurons which are then harvested and grown in cell culture (e.g. ref. 5). Immunohistochemical methods can also be used for cell identification, although markers for glutamate pathways are not as well-developed as those for some other transmitter phenotypes. Another problem of dissociated cell cultures is the difficulty in culturing neurons from adult animals (but see ref. 6). Thus, virtually all studies of glutamate receptors on neurons have used tissue from embryonic or newborn animals.

Table 1. Comparison of *in vitro* preparations for studies of glutamate receptors

Preparation	Advantages	Disadvantages
Tissue slice	● preserved synaptic connections ● adult tissue can be used	● poor access to extracellular solution. ● patch recording more difficult ● visibility of cells limited during recording
Acute dissociated neurons	● patch clamp methods easy ● adult tissue can be used ● good voltage clamp control	● metabolic disruption of cell ● loss of dendritic arbor and synaptic connections
Primary cultured neurons	● patch clamp methods easy ● concentration clamp methods can be used ● synaptic interactions can be studied	● requires embryonic or neonatal tissue ● cell identification difficult ● culture conditions can affect cell properties
Xenopus oocytes	● easy to voltage clamp ● patch clamp methods possible ● use in molecular cloning	● rapid solution changes difficult due to large cell ● appropriateness of expression and regulation is a concern

Potentially, this could bias results obtained from such studies to embryonic or neonatal receptor isoforms. This is more of a theoretical than a practical concern at the moment as no clearly defined differences between neonatal and adult animals have been reported. Improvements in cell culture techniques such as use of astrocyte feeder layers (e.g. ref. 7) or sandwich cultures (8) do permit reliable growth of neurons from animals as old as two weeks. Stable cell lines derived from tumours are relatively easy to grow in cell culture and would be of great potential benefit for studies of glutamate receptors, however, no such lines that express glutamate receptors are currently available.

Acutely dissociated CNS neurons provide a number of distinct advantages over cell cultures. In particular, cells can be taken from adult animals and used immediately in physiological experiments. Such neurons show responses evoked by L-glutamate, but several groups have noted that NMDA responses are greatly attenuated, presumably due to enzymatic digestion of extracellular components of the receptor (9). However, we have found that gentle use of papain (e.g. 5 U/ml, Worthington Biochem.) as a dissociating enzyme results in preparations of acutely dissociated neurons that have NMDA channels with typical properties, suggesting that this can be used for studies of glutamate receptors. The use of mRNA-injected *Xenopus* oocytes in studies of glutamate receptors is discussed in Section 9.

3. Pharmacology

The specificity and selectivity of agonists and antagonists available for studies of glutamate receptors has recently been reviewed (10), thus only a few general points will be addressed here. Ion channels linked to glutamate receptors were originally divided into three categories based on the selective actions of NMDA, kainic, and quisqualic acids. A structural analogue of quisqualate, AMPA (α-amino-3-hydroxy-5-methyl-4-isoxazoleproprionic acid) is more specific than quisqualate in binding studies and thus many investigators now refer to quisqualate receptors as AMPA receptors. However, as with receptor classifications for other transmitter systems, this nomenclature is undergoing continual change as more selective antagonists are developed, and until receptor structures are defined. The most conservative nomenclature combines kainate and quisqualate receptors together under the term 'non-NMDA' receptor. This reflects the inability to distinguish kainate from quisqualate responses using currently available antagonists.

Obviously, L-glutamate activates multiple receptor subtypes and thus is often not the ideal agent for use in the studies of receptor conductance mechanisms. Although this can now be overcome with the concurrent use of NMDA or non-NMDA antagonists to separate receptor subtypes activated by L-glutamate. NMDA, kainate, and quisqualate all show selective action at reasonable concentrations; however, at high concentrations each agonist may activate other glutamate receptor subtypes. For example, high concentrations of quisqualate can activate NMDA channels (11). An additional practical problem is that some commercial supplies of agonists may be contaminated with other excitatory amino acids. For example, samples of quisqualic acid from many suppliers can be contaminated with L-glutamate (12). In some cases, HPLC analysis of the samples may be necessary to verify the purity of the compounds.

A particular set of pharmacological problems apply to studies of NMDA receptors. NMDA channels are blocked by extracellular magnesium (13, 14) and require glycine for their activation (15). Thus, control of these two modulators are important for interpretation of studies of NMDA receptors. The high affinity binding of magnesium and glycine to the NMDA receptor/channel make this a particularly difficult problem as commercial salts (e.g. NaCl) often contain sufficient magnesium to affect the NMDA response, and even ultrapure water sources can contain low nanomolar concentrations of glycine sufficient to potentiate NMDA responses. This may require either analysis of the experimental solutions for magnesium and glycine contamination, or various bioassays to calibrate their concentration in the extracellular solution. The major source of extracellular glycine appears to be glial cells. In cell culture, astrocyte feeder layers release glycine into the bathing media (16), thus it is necessary to rapidly and continuously perfuse cells in order to insure that a known concentration of glycine reaches the cell membrane. Obviously these issues are even more serious in tissue slices or *in vivo* where it is difficult if not impossible to adequately remove ambient glycine and magnesium. The effects of endogenous glycine can be partly overcome now using antagonists that block the strychnine-insensitive glycine binding site on the NMDA receptor (e.g. refs 17, 18). If glycine is added to extracellular solutions in studies of NMDA receptors, it may be necessary to also add strychnine if the neurons also have strychnine-sensitive glycine receptors.

L-Glutamate can also activate second-messenger pathways via receptors that are pharmacologically distinct from 'ionotropic' receptors. The pharmacology of these receptors is less clear at the moment, although L-AP4 (L-2-amino-4-phosphonobutanoic acid) is an agonist at one such receptor in depolarizing bipolar cells in the retina (19), as well as on pre-synaptic terminals in many pathways in the central nervous system (20). Although both of these mechanisms are likely to activate G proteins in coupling to second-messenger systems, it is unclear if these two responses are identical. In addition, a 'metabotropic' glutamate receptor has been described on *Xenopus* oocytes (see below) and in cultured hippocampal neurons that couples via inositol trisphosphate to intracellular calcium stores (21). L-Glutamate, quisqualate, ibotenate, and *trans*-ACPD (*trans*-1-aminocyclopentane-1,3-dicarboxylic acid) also activate the metabotropic receptor. No antagonists for this receptor have been reported.

Overall it is now easily possible to separate NMDA, non-NMDA, and metabotropic responses using the currently available pharmacological agents. Thus, use of weak and non-selective antagonists such as GDEE (glutamic acid diethyl ester) should be avoided. The development by Honore and co-workers (22) of the quinoxalinediones has been particularly helpful. DNQX and CNQX (6,7-dinitro- and 6-cyano-7-nitro- quinoxaline-2,3-dione, respectively) and the recently developed NBQX (2,3-dihydroxy-6-nitro-7-sulphamoyl-benzo(F)quinoxaline, ref. 23) all show potent antagonism of non-NMDA receptors. However, the quinoxalinediones, especially DNQX, also act as inhibitors of the glycine site on the NMDA receptor (24). Thus to selectively block non-NMDA responses, saturating concentrations of glycine should be included in the extracellular solution.

4. Whole-cell recording

The whole-cell variant of patch-clamp recording is now in routine use in many laboratories throughout the world, and there are few specifics that apply only to glutamate-mediated responses or CNS neurons; standard information on these techniques can be found in ref. 25.

4.1 Basic issues

The essential criteria for analysing ligand-gated conductances is adequate control of voltage and space clamp errors, a problem common to both microelectrode and whole-cell voltage clamp. Voltage errors are more reliably controlled, especially on large neurons, with a discontinuous voltage clamp such as the Axoclamp-2; this allows monitoring of the actual membrane voltage. The low impedance of patch electrodes used for whole-cell clamp allows particularly rapid switching between the current-passing and voltage-measuring periods, and thus is an excellent approach for clamping large macroscopic currents. Patch-clamp amplifiers, on the other hand, clamp the pipette voltage and thus are subject to considerable errors when voltage clamp of large membrane currents is attempted. There is no perfect solution to space-clamp problems on processing-bearing neurons (see ref. 26); however, inadequate space clamp can be minimized by recognizing that the frequency components of the conductance determine the degree of error, i.e. rapid synaptic conductance changes are more subject to space-clamp error than are responses to slow applications of drugs. For a more detailed description of these problems see ref. 27. A second general problem is blocking other conductances that may contaminate the conductance mechanism under study. This involves not only adequate control of voltage so that voltage-dependent conductances are not activated, but also using appropriate pharmacological agents such as bicuculline or picrotoxin to block $GABA_A$ chloride channels, and calcium buffers to block secondarily-activate conductances such as calcium-dependent chloride channels. The composition of solutions used to do this depends on the particular example.

A major advantage of the whole-cell method is access and control of the intracellular solution. For example caesium ions are commonly used to block outward potassium currents and thus simplify analysis of L-glutamate activated inward currents. The degree of dialysis of intracellular contents is highly dependent on several factors including the size of the cell under study, the size of the pipette tip, the access-resistance of the whole-cell recording and the molecular weights of the soluble cytoplasmic components. A disadvantage of this method is the removal of important intracellular constituents, such as energy sources and second-messenger components. Although glutamate-activated channels can be recorded without adding ATP to the patch pipette, slow 'rundown' of NMDA responses does occur and has been partially overcome by inclusion of ATP in the pipette (28). NMDA responses also appear to be sensitive to changes in intracellular calcium (29). This appears to be due to a calcium-dependent component of NMDA receptor desensitization and can largely be avoided using tight intracellular buffering of calcium with high concentrations of EGTA or BAPTA (30). Several reports also suggest that kainate responses can be modified by cyclic AMP dependent phosphorylation suggesting that control of pipette contents may be critical to certain components of the kainate response (31). Although there are few whole-cell studies of glutamate-mediated second-messenger pathways, it is clear from studies of other such receptor mechanisms that careful attention to intracellular components will be necessary to adequately investigate these responses. The perforated patch method of whole-cell recording (32) which uses nystatin to punch holes in the membrane and provide electrical access without dialysing large-molecular-weight cellular proteins may be of value for such studies.

4.2 Analysis of voltage-dependent conductances

The first clue that the depolarizing action of L-glutamate resulted from activation of more than one specific ion channel came from voltage recording in cat motoneurons *in vivo* by Engberg and colleagues (33). Depolarizations evoked by NMDA receptor agonists were associated with increases in membrane resistance whereas depolarizations evoked by kainate were associated with decreases in membrane resistance. The basis of these agonist-specific changes in membrane resistance are now clear from voltage-clamp studies which demonstrated that there are two distinct conductance mechanisms activated by excitatory amino acids, a voltage-dependent conductance linked to NMDA receptors and a voltage-insensitive conductance linked to kainate/quisqualate receptors (reviewed in ref. 4). L-Glutamate activates both the voltage-dependent and voltage-independent channels and thus often shows an intermediate voltage-sensitivity in whole-cell recording. These patterns of voltage sensitivity are easily apparent in current–voltage plots generated by several different approaches, repeated pulse applications of agonist at different holding potentials, or by a series of voltage jumps or a voltage ramp in the presence and absence of agonist. The difference current represents the agonist-gated conductance. The inward current activated by NMDA has a region of negative slope conductance at membrane potentials negative to −30 mV. The voltage-sensitivity of the channel results not from a voltage-induced conformational change in the channel protein, but rather to magnesium ions entering and blocking the open channel. This effect is voltage-sensitive and increases with membrane hyperpolarization since the charged blocking ion senses the transmembrane voltage when the channel is open (13, 14). Reductions in extracellular magnesium not only increase the available conductance, but also shift the region of negative slope to more hyperpolarized membrane potentials. Despite the high affinity of the channel for magnesium, physiological levels of magnesium (1–2 mM) do not completely occlude macroscopic NMDA currents.

The onset of block of open NMDA channels by magnesium is primarily a diffusion limited process, thus studies of the kinetics of magnesium block require the time-resolution of single-channel studies (see Section 5.1). However, whole-cell recording and I–V plots of the NMDA-activated conductance have been used to show that relief of magnesium channel block can occur at extreme hyperpolarizations. Changes in extracellular calcium have complex, although predictable, effects on the NMDA-activated conductance (4). This can give rise to apparently contradictory results; for example, a decrease in magnesium and an increase in calcium can decrease the total agonist-activated conductance at some membrane potentials.

4.3 Open-channel blockers of the NMDA channel

Magnesium both enters and exits the NMDA channel rapidly. However, a number of drugs that block NMDA channels exit the channel much more slowly, and in some cases the channel can close with the drug still bound within the channel. This slow off-rate can be predicted from the equilibrium binding constants of these drugs, the highest affinity ligands such as the anticonvulsant compound MK801 [(+)-5-methyl-10,11-dihydro-5H-dibenzo [a,d]cyclohepten-5-10-imine maleate] with a $K_i \approx 4$ nM will remain in the channel for the longest time (34). Block of NMDA channels by MK801 is concentration- and use-dependent (*Figure 1*). The blocking rate of MK801, as determined by single-channel recording, is rapid (3×10^7 M^{-1} sec^{-1}), close to being diffusion limited, not voltage-dependent (between −70 and +30 mV) but concentration- and use-dependent. The unblocking rate is very slow, voltage-dependent and requires the continued presence of agonist. The unblocking rate is faster at positive membrane potentials, presumably due to 'knockout' of MK801 by outward flux of potassium or caesium ions through the channel. Block of macroscopic NMDA currents measured with whole-cell recording, however, is quite slow, the rate of onset ranging from about 8 sec at 10 μM to about 1.5 min at 100 nM. This discrepancy can be explained by a low open probability, about 0.002. While an open channel will be blocked by MK801 very quickly, each channel is open so infrequently that block of the whole-cell current, made up of hundreds or thousands of channel openings, will be greatly prolonged (34). Similar techniques for analysis of use-dependence have been reported for ketamine and phencyclidine (35, 36).

4.4. Fluctuation (noise) analysis of EAA conductances

Steady-state fluctuation analysis provides a convenient technique for determining the kinetic characteristics and conductance of glutamate-activated ion channels (e.g. 37, 38). Although the resulting channel properties are calculated rather than directly measured as in single-channel recording, a major advantage is that results can be rapidly obtained and analysed; in addition, simultaneous measurement of the whole-cell current allows a comparison of effects of drug effects on single-channel properties with antagonism of the macroscopic current. The technique is most appropriate for homogeneous populations of large conductance channels, thus results can

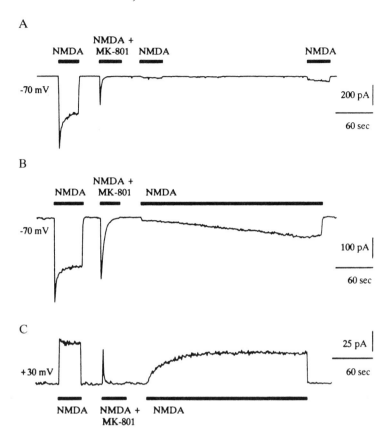

Figure 1. Recovery of NMDA responses from blockade by MK801 in whole-cell voltage-clamp recordings from cultured rat cortical neurons. A. After blocking the response to 30 μM NMDA with 10 μM MK801, the recovery of the NMDA-induced current was monitored at −70 mV by applying short pulses of 30 μM NMDA at 5 min intervals. The recovery of this cell was evaluated for 30 min but only the first two test applications of NMDA are shown. B. Following blockade by 20 μM MK801, 30 μM NMDA was applied continuously for 5 min at −70 mV. C. MK801 (10 μM) block of current elicited at +30 mV by 30 μM NMDA. All applications by flow pipe. (Reprinted with permission from Huettner and Bean, ref. 34.)

be misleading if these criteria are incorrect; such as, if significant number of subconductance levels exist. Low background noise and adequate space and voltage clamp are also essential. Low concentrations of agonist must be used in order to keep the opening probability low, and thus satisfy the general assumptions of noise analysis (39). After editing out any blocks containing spontaneous synaptic events, groups of records are grouped as epochs and an average power spectrum computed. A control spectrum obtained from baseline membrane current before drug application is usually subtracted to yield a difference spectra for agonist-induced noise. Due to increased amplifier and intrinsic membrane noise above 0.5–1 kHz, the records are low passed filtered (8-pole Butterworth). Thus noise analysis is not useful in defining high-frequency events such as rapid flickery channels block or brief channel openings (e.g. < 2 msec in duration for 1 kHz filtering).

Despite its limitations, noise analysis of NMDA channels has been useful in defining antagonists' actions (see e.g. ref. 36) and in tissue slices to demonstrate tonic activation of NMDA channels (40). Noise analysis can also be used with single-channel recording to analyse conductance levels that are too small to assess using standard single-channel analysis (38, 41).

5. Single-channel recordings

Recordings of single-channel events from excised patches of membrane have begun to define the elementary conductances linked to receptors first classified by responses of whole neurons to structural analogues of L-glutamate. This has led to molecular schemes that predict whole-cell behaviour. What has become clear is that L-glutamate and all the structural analogues thought to be specific for receptor subtypes, activate many single-channel conductances. Many of these conductance levels are activated by all of the agonists, although the different agonists activate each conductance with different probabilities.

5.1 *N*-methyl-D-aspartate channels on cultured neurons

5.1.1 Conductance levels and ion permeability

NMDA primarily gates channels with a conductance of about 50 pS (13) although smaller conductance events are almost always reported (42–45). While in the open state, 50 pS events can transiently jump to several different conductances of sizes similar to events that open from baseline (38, 42–44). The existence of these levels suggests the possibility of regulation of NMDA responses by intracellular alterations of the channel protein.

In cerebellar granule cells, 50 pS events account for 75–85% of all events activated by L-glutamate, aspartate, and NMDA while 40 and 30 pS openings comprise 10–15% and 3–7% of all conductances (38). In spinal cord neurons, a 35 pS conductance occurs less frequently than the 50 pS event and has a shorter mean open time (1.4 msec and 7.4 msec, respectively; ref. 44). The large conductances (30–50 pS) occur in bursts of openings and these bursts are clustered into periods of prolonged activity which can last for hundreds of milliseconds (43, 45). Open time distributions of the 50 pS events are best fitted by two exponentials with time-constants of about 1–3 msec and 10–15 msec (43–45).

The 50 pS conductance is the primary conductance which underlies the effects of divalent cations on whole-cell NMDA responses. Calcium is very permeable (43, 46), magnesium produces a voltage-dependent block (13, 46) and zinc and cadmium decrease its probability of opening in a voltage-independent manner (47, 48). Whole-cell recordings and experiments using fluorescent calcium indicators (see below) have shown that calcium carries a fraction of the current activated by NMDA (29, 49, 50). Single-channel recordings indicate that the 50 pS conductance (and probably events larger than about 30 pS) activated by NMDA and L-glutamate is also permeable to calcium (43, 46). Increasing extracellular calcium shifts the reversal potential to more positive values and decreases the conductance of inward currents (46). In isotonic external calcium (nominally zero external monovalent cations), the inward limb of the current voltage relation shows a conductance of about 15 pS while the reversal potential is about +35 mV (46). These measurements indicated a permeability ratio of Ca/Na of about 8 and a conductance ratio of about 0.25.

5.1.2 Single-channel analysis of NMDA channel blockers

External magnesium blocks NMDA activated channels with a voltage-dependence expected for a divalent cation. Although the mechanism of blockade has not been completely determined, it is clearly not consistent with a sequential open channel block model (13, 46) in which presence of the blocker should not change the total time the channel spends in the open state (51, 52). Instead, as the concentration of external magnesium increases, the time spent in the open state decreases. Ascher and Nowak (46) have suggested that this phenomenon could be explained by assuming that the blocked channel could close before magnesium left the pore. Recent analysis of single-channel recordings suggest that a three-state model that includes this blocked to closed transition can satisfactorily explain whole-cell responses to NMDA in the presence of physiological and higher magnesium concentrations at holding potentials between −80 mV and +60 mV (53). Analysis of single-channel events and whole-cell responses at lower magnesium concentrations down to 0.2 μM indicates that a second magnesium-independent 'blocked' state is required to explain the data.

Zinc and cadmium decrease whole-cell responses to NMDA with K_Ds of 13 and 48 μM, respectively (54). Unlike the block produced by magnesium, the degree of antagonism by zinc and cadmium is not altered by membrane potential. Single-channel recordings indicate that these divalent cations act primarily by decreasing the frequency of channel opening and the mean open time by binding to a site on the extracellular domain of the receptor. MK801 reduces both the mean open time and the number of NMDA channel openings seen in single-channel recording (34). This is consistent with the whole-cell results suggesting that MK801 rapidly blocks an open channel (thus reducing observed mean time), but then remains bound for a period exceeding the duration of the patch recording. Similar effects would be expected for other slow channel blockers.

5.2 Non-NMDA channels

The best evidence that kainate and quisqualate act through different receptors comes from single-channel recording. While they both activate a similar range of conductances, kainate primarily activates 4–5 pS events whereas quisqualate mainly activates a 8–10 pS conductance and 15–35 pS conductances (38, 41–43) although in cerebellar granule cells, kainate primarily activates 8 and 15 pS channels (38). In addition, analysis of kainate-induced noise in outside-out patches and whole-cell recordings indicates the existence of a very small conductance (140 fS) in cerebellar granule cells which may account for the majority of charge transfer in these cells (38). Whether one conductance or several underlie the fast, non-NMDA component of the synaptic current remains unknown, although if the excitatory post-synaptic current (EPSC) turns out to be limited by desensitization (see below), the rapidly desensitizing 18 and 35 pS conductances will be the prime candidates (55, 56).

5.3 Technical issues for recording of single EAA channels
5.3.1 Kinetic analysis and multichannel patches

Kinetic analysis of single-channel events is hampered by the presence of multiple channels in a membrane patch as it is often not certain whether adjacent events in time result from openings of the same channel or separate channels. Evaluation of channel kinetic schemes would ideally rely on data from patches containing only one channel. However, much can be learned using multiple channel patches, if short periods of activity can be assumed to be due to a single channel. This assumption can be met with channels that are mainly desensitized by high concentrations of agonist (57), or at low concentrations of agonist with channels that open and close repeatedly with the agonist still bound. The NMDA receptor channel fits in the second category. The K_D for L-glutamate is approximately 1 μM (58, 59), nearly 100-fold lower than the nicotinic acetylcholine receptor (AChR). If one assumes similar association rates for ACh and NMDA, then it would be expected that L-glutamate would remain bound to the NMDA receptor about 100 times longer than ACh binding to the AChR. This appears to be the case. While the lifetime of ACh on endplate channels is about 1 msec (39), 5 msec pulses of L-glutamate activate bursts of channel activity that last hundreds of milliseconds (60). In experiments in which patches are exposed to agonist for long periods (minutes), the consequence of long bound states translates into openings which are segregated into clusters of bursts. Very low agonist concentrations evoke these clusters with low probability such that they are assumed to be due to activity of a single channel. Open and closed time distributions of the clustered events then can lead to accurate predictions of kinetic models and rate-constants that link open and closed states.

5.3.2 Patch recording configurations

There have been few reports of glutamate channels in the cell-attached mode of patch clamping, reportedly due to the infrequency of activated events in these conditions. This is a significant disappointment since in cell-attached recording, the cytoplasmic face of the channel is still in contact with the normal intracellular milieu. Since it has been reported that outside-out recording can alter the kinetics of other ligand-gated channels (61), cell-attached glutamate channel recording would be of great interest. In addition, this mode of single-channel recording is potentially useful in studying regulation of channel behavior by endogenous extracellular substances.

6. Concentration clamp techniques

Just as voltage steps are required to fully evaluate the kinetics of voltage-gated conductances, concentration steps are important for kinetic analysis of ligand-gated channels. Dissociated neurons, whether acutely isolated or maintained in culture for days or weeks are ideally suited for concentration clamp techniques. Physical barriers and uptake systems which can significantly slow the arrival and decrease the final concentration of exogenous drug in slice preparations (62) are minimized in monolayer cultures, although not totally eliminated (see e.g. ref. 63). To take advantage of these assets, methods of changing external solutions have evolved to the point that complete exchanges can be obtained in tens of milliseconds with whole-cell recording (59) and in ≤ 1 millisecond with excised patches (55, 56, 60).

6.1 Whole-cell recordings

Recently, Johnson and Ascher (15) reported a technique of applying drugs to whole neurons that previously had been used only for excised patches (64) which uses streams of solutions flowing from an array of large-bore glass tubes (100–400 μm diameter). By moving the array of glass tubes, the solution from an adjacent tube rapid replaces the previous solution over the entire neuritic arbor of the neuron. This technique is so effective in exchanging solutions surrounding neurons that Johnson and Ascher (15) used it to demonstrate the requirement for glycine in NMDA receptor activation. If puffer pipettes were used to apply NMDA to a neuron bathed in medium conditioned by cultured cells (both glial and neurons) large responses were produced. However, if NMDA was applied by flow tubes, very little response was seen, suggesting that NMDA by itself is not sufficient to gate NMDA channels. Large responses could again be obtained if NMDA was applied in conditioned medium or with glycine indicating that glycine is constantly being released from surrounding cells.

To standardize and shorten the delay in switching between different flow pipes, Mayer and his colleagues have mounted flow tubes on a stepping motor drive which, in conjunction with solenoid valves to turn solutions on and off, allows switching times on the order of tens of milliseconds. The repeatability and the increased speed has allowed study of the rapid desensitization of quisqualate responses (65) and a quantitative series of experiments on the mechanism of glycine potentiation of NMDA responses (30). In the latter experiments, in the presence of low extracellular calcium and high internal calcium buffering (to avoid Ca-dependent desensitization), NMDA receptor desensitization was diminished by increasing glycine concentrations (*Figure 2*). The EC_{50} of glycine to block fast desensitization matched that found for glycine potentiation of NMDA responses (approx. 200 nM, ref 15). Despite the rapid switch of solutions possible with this system, the rate of change is dependent on the distance across which the new solution must travel before the neuron is

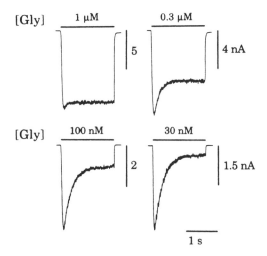

Figure 2. Concentration-dependent regulation by gylcine of NMDA receptor desensitization. Traces show the whole-cell voltage-clamp responses from cultured mouse hippocampal neurons to rapid applications of 100 μM NMDA at the glycine concentrations indicated and are plotted so that the peak amplitude of the traces is similar. Desensitization is pronounced at low concentrations of glycine, but essentially absent at high concentrations of glycine. Rapid applications were made by opening a flow pipe positioned in front of the neuron which contained agonist and simultaneously turning off the adjacent control flow pipe. (Reprinted with permission from Mayer, Vyklicky, and Clements, ref. 30.)

completely engulfed in new solution. This distance is determined by the boundary of the dendritic arbor of the neuron, usually at least 100 μm. The more compact the arbor, the faster the change can be effected. This limitation has prompted the use of outside-out patches of membrane to study the kinetics of responses to glutamate agonists.

6.2 Excised patch recording

Because the diameter of excised patches of membrane is small (approx. 1–2 μm), the external solution can be switched very rapidly. Two techniques have been used to facilitate fast changes. Tang, Dichter, and Morad (55) have used double-barrelled glass 'theta' tubing to superfuse outside-out patches with control solution from one barrel and then, by turning on flow from the second adjacent barrel, engulfing the patch in a new solution. The time required for completing this change was about 1 msec judging from junction potential generation at the tip of an open pipette. The rate of change of solution at the surface of an outside-out membrane was not determined. By using this technique they have shown that both quisqualate and L-glutamate activate what appears to be a 35 pS channel which desensitizes completely with a time-constant of 3–8 msec. Recovery from desensitization was 350 msec when quisqualate was the agonist and 100 msec with L-glutamate.

The second technique used to rapidly change the external solution at the surface of an excised patch of membrane uses a piezoelectric translator to rapidly move streams of solutions from flow pipes across the patch-pipette. Franke *et al.* (66) used this technique to activate glutamate-gated channels in crayfish muscle; channel activity began within 0.5 msec. This technique has now been used to activate glutamate-gated channels in patches from chick spinal cord neurons (56), as illustrated in *Figure 3*. Trussell and Fischbach reported fast activation kinetics of non-NMDA channels (time to peak within 1 msec, similar to the solution exchange time at the membrane surface) and slow activation of NMDA-gated channels (about 10–15 msec). On the time-scale tested, NMDA responses showed little desensitization, while the non-NMDA responses desensitized almost completely with two time-constants of 2.2 and 13.5 msec. The time-course of recovery from desensitization depended on the length of L-glutamate application. Short applications of L-glutamate (about 1 msec) resulted in recovery time-constants of 9.2 msec, while with longer application the channels entered a state of desensitization which required a much longer period for full recovery ($\tau = 83$ msec).

These studies raise the possibility that densitization may underlie the decay of the non-NMDA component of the EPSC. The best measurements of the time-constant of decay of the EPSC, however, indicate that it is 5 to 10 times faster than that of desensitization. Finkel and Redman (67) have shown that Ia afferent input to well-clamped regions of motoneurons evokes EPSCs with decay time-constants of 0.3 to 0.4 msec at 37°C). EPSCs activated by somatic

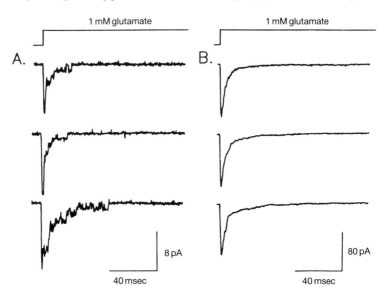

Figure 3. Responses of two different outside-out membrane patches from cultured chick spinal cord neurons of 100 msec pulses of 1 mM L-glutamate. Note the tenfold difference in sensitivity of the two patches. The top traces indicate the time of the application of agonist. (Reprinted with permission from Trussell and Fischbach, ref. 56.)

inputs on spinal cord neurons in culture (68) had decay times as fast as 0.6 msec (25°C). Concluding on the basis of these discrepancies in the decay time-constants that the mechanism of the EPSC decay is not due to desensitization may be invalid, however. If the diffusion of transmitter from the pre-synaptic terminal to post-synaptic receptors is comparable to that at the neuromuscular junction, transmitter concentration should reach its peak at the receptors in 15 µsec (69) to 300 µsec (70) whereas exogenously applied agonist, regardless of switching rate of the bulk solutions, must diffuse through an unstirred layer of approximately 1–2 µm which could take 1 msec or more (71). Given a very fast desensitization process, the relatively slow increase of exogenous agonist at the receptors due to diffusion through the unstirred layer would lead to a prolonged decay time.

A schematic of the methods we currently use for rapid solution exchanges in whole-cell and excised patch recording is shown in *Figure 4*. The piezoelectric translator is best suited to small movements (5–40 µm) and for use with excised patches.

7. Kinetics of synaptic responses

Under ideal circumstances, the kinetics of the synaptic current provide important information about the single-channel properties of the transmitter-

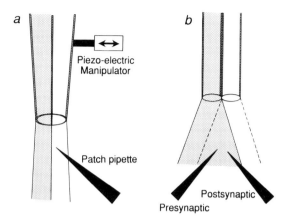

Figure 4. Rapid application methods used in our laboratories for excised patch recording (a) and whole-cell recording (b). In (a), two solutions constantly flow from a double-barrelled pipette pulled from thin-walled theta tubing (50–150 μM i.d.). The excised patch is placed within 5–15 μm of the interface of the two solutions. The solution interface is moved by piezoelectric driver attached to the delivery tube. This system allows both rapid onset and offset of solutions with time constants approaching the diffusion limit of the unstirred layer (c. 200 μsec). It also avoids the turbulence created by switching solution flows on and off. Shown in (b) is the technique we use for rapid solution changes during whole cell recording, or during recordings from pairs of synaptically-coupled neurons in cell culture. The black electrodes indicate the location of pre- and post-synaptic neurons located within a ≈ 1 mm field. Two adjacent flow pipes (i.e. 300–400 μM) are positioned near the cell(s) and one solution is switched on (grey shaded region); at solution change, the solenoid controlling barrel 1 is switched off and barrel 2 is simultaneously switched on using 5V pulses to activate the valves via a custom-built valve controller circuit. Solutions changes can be obtained with time constants of approximately 10 msec, depending on the dendritic geometry of the neurons.

gated ion channels. This is important since single-channel recordings are usually performed under equilibrium conditions using membrane patches that are likely to be extrasynaptic. Unfortunately, adequate voltage clamp of fast synaptic currents is extremely difficult; however, both voltage clamp and excised patch recording have recently provided new information concerning the properties of synaptically activated glutamate receptors.

7.1 Synaptic activation

Dissociated cell culture preparations greatly increase the ease with which simultaneous recordings can be made from two neurons that are synaptically connected. Forsythe and Westbrook (7) have shown that excitatory synapses in culture have many of the characteristics expected from the activation of both NMDA and non-NMDA receptors. EPSCs consist of a fast component and a slow component. The slow component is blocked by competitive NMDA receptor antagonists, external magnesium at hyperpolarized but not depolarized potentials, zinc at all potentials, is dependent on glycine (16) and reverses at more positive potentials in high external calcium. The early, fast component of the EPSC is not altered by these manipulations but is blocked by the non-NMDA receptor antagonist CNQX which, in the presence of high concentrations of glycine, has little effect on the slow EPSC component. In the presence of CNQX it is clear that the slow, NMDA component of the EPSC rises more slowly than the early component (60, 72), as has been shown with Schaffer collateral stimulation of hippocampal pyramidal cells (73). If the difference in activation rate of the two components cannot be accounted for by segregation of receptors at different electrotonic distances from the recording site, then either NMDA receptors activate more slowly than non-NMDA receptors or NMDA receptors are located further away from the site of transmitter release. Responses of outside out patches to rapid application of L-glutamate indicate that the activation rate of NMDA receptors is slow enough to account for the slow rise of the NMDA component of the EPSC (60). In similar experiments, the quisqualate receptor current evoked by L-glutamate activated in less than 1 msec (56). It may not be necessary to ascribe the differences in onset of the two synaptic components to differences in the activation kinetics of the two receptor/channels, however. If the receptors that produce the fast component of the EPSC were not subject to desensitization, the amplitude of the current produced could be greatly enhanced and the onset slowed. If manipulations that block densensitization can be developed (65) then this question could be addressed.

8. Optical measurements

Optical methods have a vast potential in cellular physiology as intracellular indicators are improved and new techniques such as confocal microscopy and flash photolysis come into wide usage. Optical techniques have been employed to measure the effects of glutamate receptor agonists on intracellular calcium concentration and to determine the role of intracellular calcium in regulating the efficacy of excitatory synaptic transmission. A detailed discussion of these general issues in calcium imaging can be found in Deweer and Salzberg (74).

8.1 Intracellular calcium measurements

The calcium indicators, arsenazo III and fura-2, can either be introduced into the cytoplasm by diffusion from a patch pipette in the whole-cell configuration or, in case of fura-2, applied extracellularly as the membrane permeable acetoxymethyl ester which is de-esterified and trapped intracellularly. By measuring the absorbance spectrum of arsenazo III or the ratio of fluorescent emissions at excitation wavelengths of 340 and 380 nm for fura-2, changes in intracellular calcium concentration can be detected. Using these indicators,

NMDA receptor activation has been shown to cause a large increase in intracellular calcium which is blocked by external magnesium in a voltage-dependent manner (29, 50), is dependent on extracellular calcium and blocked by the NMDA channel blocker MK801 (75). These results indicate that NMDA gated channels are permeable to calcium and describe a mechanism that can significantly increase intracellular calcium. Kainate and AMPA induce small increases in intracellular calcium by a mechanism which is not well understood but is probably due, at least in part, to activation of voltage-dependent calcium channels and the low calcium permeability of non-NMDA receptor channels (49).

Two further actions of glutamate agonists on intracellular calcium have been reported. One is mediated through a quisqualate preferring ('metabotropic') receptor which is also activated by L-glutamate and ibotentate but not AMPA, kainate, or NMDA (21, 76). This effect is not dependent on extracellular calcium or sodium, is not blocked by GAMS (γ-D-glutamylaminomethyl sulphonate), GDEE, CNQX, or L-APB and is thought to due to the release of intracellular stores of calcium, secondary to inositol

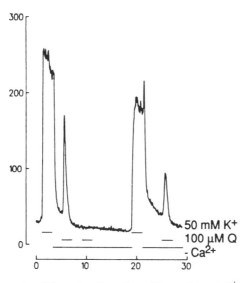

Figure 5. Depletion and refilling of quisqualate (Q)-sensitive Ca^{2+} stores. Intracellular calcium was measured in cultured mouse hippocampal neuron loaded with fura-2 by the acetoxymethyl ester procedure. The calculated intracellular Ca^{2+} activity in μM is plotted against time (in seconds). Experimental manipulations are indicated by the three lines of bars. Initially application of 50 mK K^+ to depolarize the neuron produced a large calcium transient. Intracellular Ca^{2+} returned to basal levels on removal of extracellular Ca^{2+}, but application of quisqualate produced a rapid Ca^{2+} transient; a second challenge by quisqualate was without effect. Temporary reintroduction of Ca^{2+} and depolarization-stimulated Ca^{2+} influx was sufficient to restore the response to quisqualate in Ca^{2+}-free medium. (Reprinted with permission from Murphy and Miller, ref. 76.)

trisphosphate formation. The second action of glutamate agonists is a sustained influx of calcium reported in acutely isolated hippocampal neurons which is evoked by NMDA and prevented by pre-treatment with the protein kinase inhibitor, sphingosine (77). Since hyperpolarization tended to decrease the sustained calcium influx, it was suggested that activation of this phenomenon might not be overt at the resting potential in neurons *in vivo*, but rather it might result in a long lasting potentiation of calcium influx evoked by excitatory input.

8.2 Chelators of intracellular calcium

Calcium chelators have also been used to study the effects that changes in intracellular calcium have on the efficacy of (excitatory) glutamatergic synaptic transmission. Lynch and co-workers (78) reported that intracellular injections of EGTA in concentrations sufficient to block the calcium-dependent after-hyperpolerization also blocked the induction of long-term potentiation in CA1 pyramidal cells of the hippocampus. More recently, Malenka *et al.* (79) reported similar results with the photolabile calcium chelator, nitr-5. When nitr-5 which was not loaded with calcium (and therefore had an affinity for calcium similar to that of BAPTA; see ref. 80) was injected intracellularly, LTP could not be induced although in the surrounding unperturbed cells, LTP induction was successful. In addition, step increases in intracellular calcium by photolysis of calcium-loaded nitr-5, resulted in a long-lasting enhancement of excitatory transmission without enhancement of transmission in surrounding cells. These experiments indicate that increases in intracellular calcium are required for the induction of LTP in the CA1 region of hippocampus and sufficient for at least one type of synaptic enhancement.

9. Ooctye expression of glutamate receptors

The *Xenopus* oocyte has been used increasingly in studies of ion channels due to their ability to reliably translate poly (A)+ mRNA into functional receptors and channels (81). The methods used for recording from oocytes can be found in ref. 81 and in the references cited in this section. The follicular cell layer has some neurotransmitter receptors (but no reported L-glutamate receptors), and thus is usually removed during collagenase separation of the oocytes. Nevertheless, uninjected oocytes should be checked to ensure that the responses are due to expression of foreign mRNA and not to native membrane proteins. Oocyte recording is usually performed using a two-electrode voltage clamp; however, the large capacitance of the furled membrane of the oocyte limits the time resolution of the response. Thus kinetic studies are best performed using excised patch recording from the oocyte; methods for this procedure can be found in ref. 82.

9.1 Ion channels

Early studies with poly (A)+ mRNA from whole brain regions revealed several types of glutamate responses, but not responses to NMDA. In retrospect, the lack of NMDA responses probably resulted from lack of knowledge at the time concerning the requirement for glycine. More recently, several laboratories have reported faithful translation of NMDA receptors in oocytes (83, 84). These studies have generally confirmed the expected pharmacology of native NMDA receptors (85–87), but have also produced new valuable new information [e.g. on the action of glycine (88)] primarily due to the ease of pharmacological manipulations with oocytes. The failure of RNA fractionation to identify single fractions that translate NMDA receptors (89, 90) suggests multiple subunits may be involved, consistent with the structure of other ligand-gated ion channels. Of course, the major purpose in the oocyte translation system is to obtain molecular clones of glutamate receptors. Hollman and co-workers (91) have recently reported the first clone which codes for a functional EAA receptor, a kainate receptor obtained from rat brain.

9.2 Metabotropic glutamate receptor

Many transmitter receptors which couple to G proteins can be expressed in oocytes (92). In the oocyte, receptor activation results in production of inositol trisphosphate, release of intracellular calcium and opening of a Ca-dependent chloride channel. This expression system has been useful in studies of the metabotropic glutamate receptor (93) and should provide a useful approach to molecular cloning of the G protein coupled glutamate receptors. The chloride equilibrium potential in oocytes is near −20 mV, thus the metabotropic receptor evokes an oscillating inward chloride current at negative membrane potentials which is easily recorded under voltage clamp. A similar response has been seen in cultured hippocampal cells (see Section 8.1).

10. Future directions

Studies of central excitatory synapses and glutamate receptors will benefit greatly from the molecular studies of receptors and synaptically-related proteins that are just beginning to unfold. This will allow much better definition of the types and functional properties of glutamate ion channels and G protein coupled receptors. This should also provide a wide variety of biochemical and anatomical probes for studies of central excitatory synaptic mechanisms. In addition new preparations and techniques such as the thin-slice and 'non-invasive' optical techniques should continue to narrow the gap between our understanding of cellular neurophysiology and the function of the intact system.

Acknowledgements

We thank the Medical Research Foundation of Oregon, the McKnight Foundation and Public Health Service (NIH, ADAMHA) for support of work in our laboratories.

References

1. Dingledine, R. (1984). *Brain Slices*. Plenum Press, New York.
2. Edwards, F. A. Konnerth, A., Sakmann, B., and Takahashi, T. (1989). *Pflügers Archiv*, **414**, 600.
3. Hestrin, S., Nicoll, R. A., Perkel, D. J., and Sah, P. (1990). *Journal of Physiology (London)*, **422**, 203.
4. Mayer, M. L. and Westbrook, G. L. (1987). *Progress in Neurobiology*, **28**, 197.
5. Huettner, J. E. and Baughman, R. W. (1986). *Journal of Neuroscience*, **8**, 160.
6. Lindsay, R. M. Evison, C. J., and Winter, J. (1990). In *Cellular Neurobiology: A Practical Approach* (ed. J. E. Chad and H. W. Wheal). IRL Press at Oxford University Press, Oxford.
7. Forsythe, I. D. and Westbrook, G. L. (1988). *Journal of Physiology (London)*, **396**, 515.
8. Bartlett, W. P. and Banker, G. A. (1984). *Journal of Neuroscience*, **4**, 1944.
9. Akaike, N., Kaneda, N., Hor, N. and Krishtal, O. A. (1988). *Neuroscience Letters*, **87**, 75.
10. Watkins, J. C., Krogsgaard-Larsen, P., and Honore, T. (1990). *Trends in Pharmacological Science*, **11**, 25.
11. Grudt, T. J. and Jahr, C. E. (1990). *Molecular Pharmacology*, **37**, 477.
12. Cha, J. J., Hollingsworth, Z. R., Greenamyre, T., and Young, A. B. (1989). *Journal of Neuroscience Methods*, **27**, 143.
13. Nowak, L., Bregestovski, P., Ascher, P., Herbet, A., and Prochiantz, A. (1984). *Nature*, **307**, 462.
14. Mayer, M. L., Westbrook, G. L., and Guthrie, P. B. (1984). *Nature*, **309**, 261.
15. Johnson, J. W. and Ascher, P. (1987). *Nature*, **325**, 529.
16. Forsythe, I. E., Westbrook, G. L., and Mayer, M. L. (1988). *Journal of Neuroscience*, **8**, 3733.
17. Kemp, J. A., Foster, A. C., Leeson, P. D., Priestley, T., Tridgett, R., Iversen, L. L., and Woodruff, G. N. (1988). *Proceedings of the National Academy of Sciences of the USA*, **85**, 6547.
18. Heuttner, J. E. (1989). *Science*, **243**, 1611.
19. Nawy, S. and Jahr, C. E. (1990). *Nature*, **346**, 269.
20. Forsythe, I. D. and Clements, J. D. (1990). *Journal of Physiology (London)*, **429**, 1.
21. Murphy, S. N. and Miller, R. J. (1989). *Molecular Pharmacology*, **35**, 671.
22. Honore, T. S., Davies, S. N., Drejer, J., Fletcher, E. J., Jacobsen, P., Lodge, D. and Nielsen, F. E. (1988). *Science*, **241**, 701.
23. Sheardown, M. J., Nielsen, E. O., Hansen, A. J., Jacobsen, P., and Honore, T. (1990). *Science*, **247**, 571.

24. Birch, P. J., Grossman, C. J., and Hayes, A. G. (1988). *European Journal of Pharmacology*, **156**, 177.
25. Neher, E. and Sakmann, B. (1983). *Single-Channel Recording*, Plenum Press, New York.
26. Turner, D., Stockley, E., and Wheal, H. W. (1990). In *Cellular Neurobiology: A Practical Approach* (ed. J. E. Chad and H. W. Wheal). IRL Press at Oxford University Press.
27. Johnston, D. and Brown, T. H. (1983). *Journal of Neurophysiology*, **50**, 464.
28. MacDonald, J. F., Mody, I., and Salter, M. W. (1989). *Journal of Physiology (London)*, **414**, 17.
29. Mayer, M. L., MacDermott, A. B., Westbrook, G. L., Smith, S. J., and Barker, J. L. (1987). *Journal of Neuroscience*, **7**, 3230.
30. Mayer, M. L., Vyklicky, L. and Clements, J. (1989). *Nature*, **338**, 425.
31. Knapp, A. G. and Dowling, J. E. (1987). *Nature*, 325, 437.
32. Horn, R. and Marty, A. (1988). *Journal of General Physiology*, **92**, 145.
33. Engberg, I., Flatman, J. A., and Lambert, J. D. C. (1979). *Journal of Physiology (London)*, **288**. 227.
34. Heuttner, J. E. and Bean, B. P. (1988). *Proceedings of the National Academy of Sciences of the USA*, **85**, 1307.
35. MacDonald, J. F. Miljkovic, Z., and Pennefather, P. (1987). *Journal of Neurophysiology*, **58**, 251.
36. Mayer, M. L., Westbrook, G. L., and Vyklicky, L. Jr. (1988). *Journal of Neurophysiology*, **60**, 645.
37. Westbrook, G. L., Mayer, M. L., Namboodiri, M. A. A., and Neale, J. H. (1986). *Journal of Neuroscience*, **6**, 3385.
38. Cull-Candy, S. G., Howe, J. R., and Ogden, D. C. (1988). *Journal of Physiology (London)*, **235**, 655.
39. Anderson, C. R. and Stevens, C. F. (1973). *Journal of Physiology (London)*, **235**, 655.
40. Sah, P., Hestrin, S. and Nicoll, R. A. (1989). *Science*, **246**, 815.
41. Ascher, P. and Nowak, L. (1988). *Journal of Physiology (London)*, **399**, 227.
42. Cull-Candy, S. G. and Usowicz, M.M. (1987). *Nature*, **325**, 525.
43. Jahr, C. E. and Stevens, C. F. (1987). *Nature*, **325**, 522.
44. Ascher, P. Bregestovski, P., and Nowak, L. (1988). *Journal of Physiology (London)*, **399**, 207.
45. Howe, J. R., Colquhoun, D., and Cull-Candy, S. G. (1988). *Proceedings of the Royal Society, London*, **B233**, 407.
46. Ascher, P. and Nowak, L. (1988). *Journal of Physiology (London)*, **399**, 247.
47. Christine, C. W. and Choi, D. W. (1990). *Journal of Neuroscience*, **10**, 108.
48. Legendre, P. and Westbrook, G. L. (1990). *Journal of Physiology (London)*, **429**, 429.
49. Mayer, M. L. and Westbrook, G. L. (1987). *Journal of Physiology (London)*, **394**, 501.
50. MacDermott, A. B., Mayer, M. L., Westbrook, G. L., Smith, S. J., and Barker, J. L. (1986). *Nature*, **321**, 519.
51. Neher, E. and Steinbach, J. H. (1978). *Journal of Physiology (London)*, **277**, 153.
52. Neher, E. (1983). *Journal of Physiology (London)*, **339**, 663.
53. Jahr, C. E. and Stevens, C. F. (1990). *Journal of Neuroscience*, **10**, 3178.

54. Mayer, M. L., Westbrook, G. L., and Vyklicky, L. (1989). *Journal of Physiology (London)*, **415**, 329.
55. Tang, C. M., Dichter, M., and Morad, M. (1989). *Science*, **243**, 1474.
56. Trussell, L. O. and Fischbach, G. D. (1989). *Neuron*, **3**, 209.
57. Colquhoun, D. and Ogden, D. (1988). *Journal of Physiology (London)*, **395**, 131.
58. Olverman, H. J., Jones, A. W., and Watkins, J. C. (1984). *Nature*, **307**, 460.
59. Patneau, D. K. and Mayer, M. L. (1990). *Journal of Neuroscience*, **10**, 2385.
60. Lester, R. A. J., Clements, J. D., Westbrook, G. L., and Jahr, C. E. (1990). *Nature*, **346**, 565.
61. Trautmann, A. and Seigelbaum, S. A. (1983). In *Single-Channel Recording* (ed. E. Neher and B. Sakmann), p. 473. Plenum Press, New York.
62. Garthwaite, J. (1985). *British Journal of Pharmacology*, **85**, 297.
63. Sather, W., Johnson, J. W., Henderson, G., and Ascher, P. (1990). *Neuron*, **4**, 725.
64. Yellen, G. (1982). *Nature*, **296**, 357.
65. Mayer, M. L. and Vyklicky, L. (1989). *Proceedings of the National Academy of Sciences of the USA*, **86**, 1411.
66. Franke, C., Hatt, H., and Dudel, J. (1987). *Neuroscience Letters*, **77**, 199.
67. Finkel, A. S. and Redman, S. J. (1983). *Journal of Physiology (London)*, **342**, 615.
68. Nelson, P. G., Pun, R. Y. K., and Westbrook, G. L. (1986). *Journal of Physiology (London)*, **372**, 169.
69. Matthews-Bellinger, J., and Salpeter, M. (1978). *Journal of Physiology (London)*, **279**, 197.
70. Gage, P. W. and McBurney, R. N. (1975). *Journal of Physiology (London)*, **244**, 385.
71. Brett, R. S., Dilger, J. P., Adams, P. R., and Lancaster, B. (1986). *Biophysical Journal*, **50**, 987.
72. Westbrook, G. L. and Jahr, C. E. (1989). *Seminars in Neuroscience*, **1**, 103.
73. Collingridge, G. L., Herron, C. E., and Lester, R. A. J. (1988). *Journal of Physiology (London)*, **399**, 283.
74. Deweer, P. and Salzberg, B. M. (1986). *Optical Methods in Cell Physiology*. Society of General Physiology and Wiley-Interscience, New York.
75. Murphy, S. N., Thayer, S. A., and Miller, R. J. (1987). *Journal of Neuroscience*, **7**, 4145.
76. Murphy, S. N. and Miller, R. J. (1988). *Proceedings of the National Academy of Sciences of the USA*, **85**, 8737.
77. Connor, J. A., Wadman, W. J., Hockberger, P. E., and Wong, R. K. S. (1988). *Science*, **240**, 649.
78. Lynch, G., Larson, J., Kelso, S., Barrionuevo, G., and Schlotter, F. (1983). *Nature*, **305**, 719.
79. Malenka, R. C., Kauer, J. A., Zucker, R. S., and Nicoll, R. A. (1988). *Science*, **242**, 81.
80. Tsien, R. Y. and Zucker, R. S. (1986). *Biophysical Journal*, **50**, 843.
81. Dascal, N. (1987). *CRC Critical Reviews in Biochemistry*, **22**, 317.
82. Methfessel, C., Witzemann, V., Takahashi, T., Mishina, M., Numa, S., and Sakmann, B., (1986). *Pflügers Archiv*, **407**, 577.

83. Verdoorn, T. A., Kleckner, N. W., and Dingledine, R. (1987). *Science*, **238**, 1114.
84. Kushner, L., Lerma, J., Zukin, R. S., and Bennett, M. V. L. (1988). *Proceedings of the National Academy of Sciences of the USA*, **85**, 3250.
85. Verdoorn, T. A. and Dingledine, R. (1988). *Molecular Pharmacology*, **34**, 298.
86. Verdoorn, T. A. and Dingledine, R. (1989). *Molecular Pharmacology*, **35**, 360.
87. Lerma, J., Kushner, L., Zukin, R. S., and Bennett, M. V. L. (1989). *Proceedings of the National Academy of Sciences of the USA*, **86**, 2083.
88. Kleckner, N. W. and Dingledine, R. (1988). *Science*, **241**, 835.
89. Sumikawa, K., Parker, I., and Miledi, R. (1984). *Proceedings of the National Academy of Sciences of the USA*, **81**, 7994.
90. Fong, T. M., Davidson, N., and Lester, H. A. (1988). *Synapse*, **6**, 657.
91. Hollmann, M., O'Shea-Greenfield, A., Rogers, S. W., and Heinemann, S. (1989). *Nature*, **342**, 643.
92. Snutch, T. P. (1988). *Trends in Neuroscience*, **11**, 250.
93. Sugiyama, H., Ito, I., and Hironi, C. (1987). *Nature*, **325**, 531.

4

Identifying and characterizing stretch-activated ion channels

CHRISTIAN ERXLEBEN, JOACHIM UBL, and
HANS-ALBERT KOLB

1. Introduction

1.1 Stretch-activated channels: an overview

The first observation of stretch-activated (SA) channels was probably accidental. It is a common practice to use moderate suction on the tubing connected to the patch-pipette to facilitate the formation of a gigaohm seal. The first reports of SA-channels were those of Guharay and Sachs (1984) (1) on chick myotubes and of Brehm et al. (1984) (2) on embryonic *Xenopus* muscle. Subsequently, SA-channels have been found in a variety of biological systems, ranging from bacteria, yeast, and plant cells, over molluscan heart muscle to vertebrate oocytes, lens epithelia, endothelial cell lines and kidney cells (for reviews see: refs 3, 4 and 39). The most recent reports also include K^+-selective SA-channels in growth cones of neurons, where they co-exist with stretch-inactivated channels (5), in human fibroblasts (6), embryos of fresh-water fish (7) and in stretch-receptor neurons of the crayfish (8, 9).

This article mainly deals with two aspects of SA-channels. First, by means of data from SA-channels in a mechanoreceptor, we will give practical guidelines and criteria for the analysis of a stretch-sensitive channel which is activated by suction. Second, in respect to the physiological significance of SA-channels in non-mechanoreceptors, we will evaluate the evidence for their involvement in cell volume regulation. By means of data from our own work on opossum kidney cells (10, 11, 12), we will develop criteria that will be needed to clarify the role of SA-channels in volume regulation.

1.2 Physiological function of stretch-activated channels

Surprisingly, SA-channels have mainly been observed in various animal cells which are obviously not specific mechanosensors. Therefore, the question arises as to the significance for the wide distribution of this specific channel type. Since the ability of volume regulation is a basic function of all animal cells, it has been speculated that SA-channels might be involved as osmosensors in this regulatory mechanism (see Section 5.2).

Another function of SA-channels might be an involvement in the regulation of cell growth. This is suggested by their presence in oocytes (13, 14). Insight into the mechanisms of how stretch-channels can regulate cell growth comes from observations on growth cones of cultured snail neurons. Here K^+-selective SA- and stretch-inactivated channels were found to have different ranges of activation/inactivation (5). Thus, current through the stretch channels will be minimal at some intermediate value of membrane tension. Because of the K^+-selectivity of these channels, this will lead to depolarization of the cell and, conversely, lower the threshold for excitation and concomitant influx of Ca^{2+}. Ca^{2+}-influx has been shown to be involved in the regulation of growth cone motility (15, 16). Further evidence comes from fresh-water fish embryos in which oscillatory changes in resting membrane potential are thought to result from periodic changes of membrane tension, which in turn regulate K^+-selective SA-channels (7).

Finally, SA-channels have been postulated to mediate the first step of signal transduction of mechano-sensory cells. In specialized, highly sensitive sensors, such as hair cells of the auditory system or arthropod sensilla, mechanical force-amplifying structures are used to concentrate stress towards the transducing membrane [for reviews, see Howard *et al.* (17), French (18)]. Consequently, the transducing membrane is not easily accessible for single-channel recordings and measurements of mechanically activated single-channel currents have been restricted to whole-cell recordings from vestibular hair cells (19). A direct demonstration of the presence of SA-channels in a mechanoreceptor comes from the abdominal stretch receptor neuron of crayfish (8, 9). Here it is possible to record single-channel activity from the transducing membrane at the dendrites (see Section 5.1).

2. Methods
2.1 Recording techniques and analysis of channel fluctuations

The technique of single-channel recording is well established by now, and the reader is referred to the paper of Hamill *et al.* (20), which describes the giga-seal method along with the basic design of the recording system. For a more comprehensive text which deals not only with theoretical but also with practical aspects of single-channel recording techniques *Single-Channel Recording* (21) is recommended. Hardware for acquisition of single-channel data along with the appropriate software for analysis is commercially available. In the following, relevant points will be mentioned that are of specific interest for the analysis of single-channel data from SA-channels.

As a first step in any quantitative analysis of single-channel data, channel activity, i.e. the probability of the channel being open (P_o), is determined as a function of parameters such as membrane potential, temperature, concentration of an agonist or channel blocker or, in case of the SA-channel, as a function of membrane tension. The most straightforward way to determine P_o is to integrate channel activity above the baseline over a period of time, and to divide the integral by the recording time. It is essential for this analysis that the channel activity is stationary. As a result, a current proportional to the mean number of open channels ($n \cdot P_o$) is obtained. If the number of channels and their current amplitude are known, the single-channel open probability, P_o, is then obtained by dividing by the single-channel current and the number of channels in the patch; n is usually taken as the maximum number of simultaneous channel openings that can be observed at the highest level of activity. Generally, it is assumed that the channels open and close independently. A more reliable approach for the estimation of the actual number of channels in the membrane patch uses the binomial distribution (22). The probability that out of n channels j ($1 \leq j \leq n$) are simultaneously open is determined for all j and compared with the expectation of the corresponding binomial distribution. This comparison yields an estimate for the open probability of an individual channel. The 'integration over baseline' method has the advantage that no prior knowledge about the kinetics of the channel, i.e. the existence of conductance substates, is required. A disadvantage of the method is that it is very sensitive to incorrect estimates of the baseline. In particular, at low levels of channel activity, an incorrect estimate of the baseline can result in significant errors in P_o. This problem is aggravated at high levels of suction where a shift in the baseline and increase in baseline fluctuations can often be observed. It therefore seems preferable to measure all channel openings individually and total the open times in order to calculate the time integral. Usually, the 50% amplitude level is used as threshold for the detection of channel openings and closures. Besides being more tedious, this method requires that the single-channel current amplitudes are clearly defined. This practically limits the analysis to patches with only a few active channels and will be a problem if the channels have multiple conductance states.

2.2 Hydrostatic pressure and membrane tension

Most knowledge about the elastic properties of biological membranes has been obtained from studies on red blood cell membranes and lipid vesicles of variable lipid composition. The elastic properties are characterized by the elastic area compressibility modulus (k_s). This elastic constant characterizes the membrane to area expansion or compression and is about 4×10^4 times larger than the elastic modulus of shear rigidity (23, 24). For k_s a total value of about 290 dyn/cm is determined from red cells at 25°C. The cytoskeleton contributes to this value by 193 dyn/cm (24). On red blood cells and lipid vesicles the maximal fractional area expansion which is required to produce lysis or disruption of the lipid bilayer is between 2 and 4%. This yields (see equation 2 of Section 3.1) for an isotropic membrane tension a maximal value of 10–12 dyn/cm.

2.2.1 Channel activation by pressure

SA-channels are usually stimulated by applying pressure to the patch-pipette within a range of $\pm$ 10 cm Hg ($= \pm 1.33 \times 10^5$ dyn/cm^2) seeming about maximal for biological membranes. This is most simply achieved by connecting a small volume syringe (i.e. 100 µl Hamilton, Cotati, California, USA) to the electrode holder. As long as one is only interested in the steady state behaviour of the channels, the syringe can be operated by hand. Pressure can be monitored in several ways: (a) by a simple mercury manometer or (b) by a pressure transducer in connection with a suitable bridge circuit. As a transducer, a sensor commonly used for measuring blood pressure (i.e. P23 Statham, Oxnard, California, USA), or, alternatively, an inexpensive semi-conductor pressure sensor (i.e. KPY 35R, Siemens, Germany), can be used. A complete pressure monitor with digital readout which is based on a semi-conductor sensor is also available (i.e. PMO15 WPI, New Haven, Connecticut, USA). If the transducer/syringe part of the system is filled with fluid, the response time is improved over a purely air filled system.

2.2.2 Sources of artefacts

An almost trivial explanation for how the increase in channel activity can result from suction on the patch-pipette is that more membrane is drawn into the pipette upon increasing suction. Such an artefact should be suspected if the increase in channel activity is not reversible and is more likely to happen when only a low resistance seal can be obtained. A change in membrane area can be detected by monitoring the membrane capacity during the course of an experiment. If the capacitance cancellation network of the patch clamp amplifier is used, the change in capacitance can be simply read from the dials of the instrument. It could also be argued that suction produces defects in the structure of the lipid bilayer which yield an increase of the current across the membrane patch. But it could be demonstrated for pure lipid bilayers that defects in the lipid bilayer structure yield current fluctuations of random amplitude (25). Therefore, consecutive openings of several channels of one population should appear in a step-wise current change of equal size.

3. Theory

3.1 Pressure and voltage-dependence of the SA-channel open probability

The channel open probability, P_o, increases with about the square of the applied pressure (p). This relationship has been determined for SA-channels in chick muscle, frog lens, stretch-receptor of crayfish, opossum kidney cells, and spheroblasts of *Escherichia coli* (see *Table 1*). Due to the quadratic dependence, the open probability should be independent of the polarity of the applied pressure whether suction or positive pressure is applied to the membrane patch. A theoretical explanation for this pressure dependence is given by Guharay and Sachs (1). The essential assumptions of this theoretical approach are:

(a) The shape of the membrane within the tip of the patch pipette is considered to be hemispherical with diameter (d).

(b) The membrane is homogeneous and isotropic in respect to the surface membrane tension (δ). Then Laplace's law can be used:

$$\delta = p \cdot d/4. \quad [1]$$

(c) The applied pressure causes only a weak deformation. Therefore the Hooks's relation can be applied:

$$\delta = k_s \cdot \Delta A/A \quad [2]$$

where $\Delta A/A$ is the relative change in the membrane area A and k_s the elastic area modulus of stretching with the dimension of energy per unit area (dyn/cm). Then the elastic energy of stretching per unit area (ΔW) must be a quadratic function of $\Delta A/A$ (23):

$$\Delta W = 1/2 \, k_s \cdot (\Delta A/A)^2. \quad [3]$$

(d) k_s describes the two-dimensional elastic properties of the channel embedded in the membrane. The pressure-induced transfer of free energy (ΔG) to an ion channel with a cross-section of πr^2, which yields a transition from the closed to the open channel configuration, is given by:

$$\Delta G = f \cdot (\pi r^2) \cdot \Delta W \quad [4]$$

where f ($f \leq 1$) denotes the fraction of free energy which is necessary for a conformational change from the closed to the open channel configuration. Using the Eyring rate theory, the pressure-dependence of the corresponding rate-constant $k(p)$ is given by:

$$k(p) = k(0) \cdot \exp(-\Delta G/kT) \quad [5]$$

where $k(0)$ is the rate-constant if no external pressure is applied to the patch-pipette. kT is the product of Boltzmann's constant and absolute temperature. Using the equations 1–5 it follows:

$$\Delta G = \theta \cdot p^2 \quad [6]$$

and for the pressure sensitivity:

$$\theta = f \cdot \pi \cdot (rd)^2/(32 \cdot k_s \cdot kT). \quad [7]$$

(e) The channel kinetics has been described by a linear sequential chemical reaction scheme of three closed states and one open channel state. A critical assumption is that the transition rates to channel closed states are pressure independent, which seems to hold true for SA-channels in chick muscle (1), but is not found for SA-channels in stretch-receptors of crayfish (9). The corresponding pressure dependent rate constant is the rate limiting step of the linear sequential chemical reaction. In a first approximation the steady-state channel kinetics can be described by a monomolecular chemical reaction and the corresponding channel open probability can be written as:

$$P_o(p) = 1/[1 + K \cdot \exp(-\theta \cdot p^2)] \qquad [8]$$

The pressure-independent constant K can be determined by the relation:

$$K = \exp(\theta \cdot \bar{p}^2) \qquad [9]$$

where $\bar{p}$ is the pressure for half-maximal activity ($P_o = 0.5$).

The voltage-dependent gating of channel opening appears to be independent of the applied pressure and affects only the pressure-dependent rate-constant (see above). Following the Eyring rate theory equation 8 can be extended:

$$P_o(p,V) = 1/[1 + K \cdot \exp(-\theta \cdot p^2 + \alpha \cdot V)] \qquad [10]$$

K is a constant which can be derived from measuring P_o if no external pressure is applied to the patch-pipette, and in the absence of an electrical field across the membrane (the pipette potential is held at 0 mV); α denotes the voltage sensitivity in mV^{-1}.

4. Comparison of pressure- and voltage-sensitivity of SA-channels in different cell types

According to equations 9 and 10, SA-channels can be characterized by the basic parameters of pressure- and voltage-sensitivity and the value of half-maximal activation. *Table 1* shows these values for several types of cells. Considering the diversity of organisms from which the cells were taken, it is surprising that the pressure sensitivity for chick muscle, frog lens, opossum kidney cells, as well as for *E. coli*, differ only by a factor of up to three. Also, the corresponding pressure values for which the open probability is half-maximal are found to be similar. Using equation 7, k_s could be estimated if the channel area within the membrane was known, which is affected by the transfer of free energy for channel opening (for discussion see Sachs, ref. 4). On the other hand, k_s of the membrane patch could be estimated from measurements of the minimal suction pressure yielding membrane disruption and the relative change in the membrane area (equation 2). Membrane patches generally break at a negative pressure of 7–10 cmHg (1 cmHg = 1.33 × 10^4 dyn/cm^2). Using patch-pipettes with tip diameters between 1 and 2 µm and equation 1, this would result in membrane tensions of 2–7 dyn/cm. With a value of $\Delta A/A \sim 0.03$ a k_s in the range of 70–100 dyn/cm is derived. The corresponding change in the volume (V_1 to V_2) of the hemispherical membrane patch can be derived from equation 2:

$$V_2/V_1 = (1 + \delta/k_s)^{3/2}. \qquad [12]$$

Table 1. Conductance and pressure properties of SA-channels in various cell membranes. The pressure-dependent open probability was fitted by equation 8. θ denotes the pressure sensitivity. The values in brackets denote the corresponding values for cells which were incubated in bath media containing 50 μM cytochalasin B. α is the voltage-sensitivity (equation 10). For further explanation see Section 3.1. The data denoted by * were derived by estimation or extrapolation from the figures of the corresponding reports, respectively. Values of half-maximal pressure are presented for a pipette potential of 0 mV, i.e. at the cells' membrane resting potentials. In opossum kidney cells and stretch-receptor crayfish, the observed two populations of SA-channels are denoted by I and II, respectively.

Cell	Single-channel conductance (pS)	α/mV^{-1}	$\theta/\text{cm}^{-2}\text{Hg}$	Half-maximal pressure (mmHg)	Refs
Chick muscle	35/Na 70/K	0.01	0.14 (4.34)	$\geq 70^*(12)^*$	1, 26
Frog lens	28/Na 30/K		0.5	$\sim 38^*$	27
Crayfish I	50/Na 71/K	0.012	1.7	~ 13	9
Stretch-receptor II	44/Na		0.2	> 50	
Opossum I	30/Na 22/K	0.021	0.14 (1.3)	$\sim (\sim 15)$	10–12
kidney II	42/Na		0.07	$>> 60$	
E. coli	970/Cl	0.067	0.33^*	$\sim 5^*$	28

Using a k_s of 290 dyn/cm (24) only a volume increase of about 1% is required to produce a membrane tension of 6 dyn/cm.

Table 1 also shows that the rate-constant k increases with depolarization. The corresponding voltage-sensitivity constants (α) are given in Table 1 as well; k increases with depolarization and the corresponding values for α are within a narrow range. Therefore, it appears reasonable to assume that the pressure- and voltage-dependent gating mechanisms are similar for the different cell types.

5. Examples of SA-channels and their proposed physiological function.

5.1 SA-channels in abdominal stretch-receptor of crayfish

Two types of SA channels have been described in the stretch-receptor preparation. They have similar conductance properties, but differ in their voltage range of activation and sensitivity to membrane tension (8, 9). In the following, the data of one type is discussed. The other type shows strong negative voltage-dependence of P_o, and half-maximal activation by pressure occurs above 50 mmHg. For comparison, the kinetic parameters of the channel are included in Table 1 (type II).

The preparation consists of fast and slow adapting sensory neurons along with the corresponding muscle fibres. In order to have access to the membrane it was necessary to digest the connective tissue with protease (Sigma, protease type XIV). For this purpose a patch electrode with a large opening of about 25 μm was filled with the normal bath solution (van Harreveld saline with: 195 mM NaCl, 13.5 mM $CaCl_2$, 2.6 mM $MgCl_2$, 5.4 mM KCl and 10 mM Tris with the pH adjusted to 7.4 with NaOH) containing 10 mg/ml protease. The pipette was then positioned on to the receptor neuron. An exposure of 5 to 10 min, depending on the extent of connective tissue, was sufficient to reliably obtain gigaohm seals. This method of local enzyme treatment, which should be suitable for other semi-intact preparations, was chosen over enzymatic dispersion, since it preserves the integrity of the preparation.

5.1.1 Relationship between open probability and suction

Figure 1 shows recordings from a cell-attached patch with single-channel currents from the stretch receptor neuron. At the cell's resting potential, and

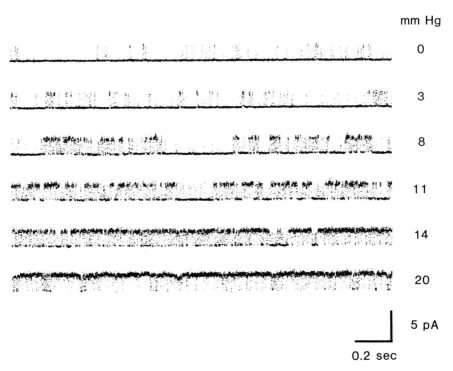

Figure 1. Unitary currents through a stretch-activated channel in the abdominal stretch-receptor neuron of the crayfish. Cell-attached patch at the cell's membrane resting potential with physiological saline in the pipette. Non-consecutive traces during constant suction levels as indicated. Data were filtered at 1 KHz and sampled at 5 KHz; upward deflection corresponds to inward current.

in the absence of suction, only a few openings of an inward current passing channel can be observed. Channel activity increases in a non-linear fashion, while membrane tension is increased by applying negative pressure to the patch electrode. During constant suction mean channel activity also remained constant. This was determined by comparing the time average of one-second segments during periods of sustained suction.

At a suction level of about 15 mmHg, the channel is open for 50%, and above some 20 mmHg, it is open most of the time. The absence of any double openings indicates that there is only one active channel in the patch. Thus, the absolute open probability for the channel as a function of the applied negative pressure can be obtained. For this purpose single-channel activity was recorded at different levels of suction. The sample time for each episode was roughly proportional to $1/P_o$ in order to provide a comparable number of observations; P_o was then determined by summation of all channel open times. The fraction of time that the channel spends in the open state as a function of pressure is plotted in *Figure 2*. Quantitatively, the steady state relationship between suction and P_o can be described by a Boltzmann curve, similar to equation 8 derived above. It is of the form:

$$P_o = 1/[1 + K \cdot \exp(-\theta \cdot (p - p_0)^2)]. \qquad [12]$$

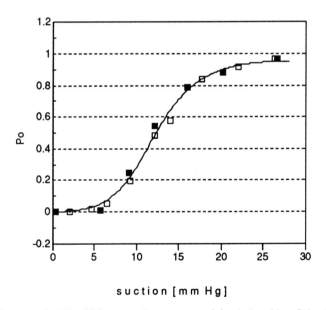

Figure 2. Open probability (P_o) vs. suction pressure (p) relationship of the SA-channel from the stretch receptor neuron. P_o is plotted vs. negative pressure applied to the pipette. Open and closed squares are data from two series of measurements from the same cell-attached patch (same as *Figure 3*). The line is a fit to the datapoints with equation 12. Parameters of the fit are: $K = 0.05$, $\theta = 1.13$ (cmHg^{-2}) and $P_o = 2.87$ (cmHg).

K is a pressure insensitive term and θ the sensitivity. p denotes the applied negative pressure in cmHg and p_0 is an offset pressure term (see also ref. 4). p_0 is introduced to account for the pre-conditioning of the membrane patch which is caused by the formation of the seal. At least in theory, p_0 could be determined according to equation 1 by comparing P_o in the absence of pressure for patch-pipettes of different diameter. Equation 12 assumes a squared dependence of the energy used for channel gating and the suction applied to the pipette. It should be noted, however, that the data show sufficient scatter to be equally well fitted by a Boltzmann distribution with a linear dependence between P_o and p. The plot contains data from two series of measurements, each with values combined from ascending and descending pressure series. The fact that the points from both runs agree quite well indicates: (a) that the effect of membrane tension on the channel is reversible, and (b) a long-term stationary behaviour of the channel. Non-stationary behaviour of SA-channels is observed in several other systems (4, 27) where it impedes quantitative analysis.

5.1.2 Current–voltage relationship and selectivity

Like most other SA-channels, the channel in the crayfish stretch-receptor preparation is a non-selective cation channel. This is evident from the current(I)–voltage(V) relation of the single-channel current amplitudes from a cell-attached patch with normal van Harreveld saline in the pipette (*Figure 3*). (a) The *I–V* curve is linear over a wide range of potentials. (b) Current reversal is close to 0 mV (assuming a cell resting potential of 65 mV) as expected for a channel that passes mainly sodium and potassium ions under physiological conditions. Similar current–voltage relations were obtained from experiments where the patch electrodes contained isosmotic solutions with either K^+ or Ca^{2+} ions as charge carrier. The average slope conductance for the stretch-receptor SA-channels was found to be 71 pS for K^+, 50 pS for Na^+, and 23 pS for Ca^{2+} (9).

5.2 Channel-activation during volume regulation

Volume regulation of cells pertains to the ability of cells to regain their physiological cell volume after exposure to a medium of anisotonic composition. This regulatory function has mainly been studied during the cellular response to a constantly applied hypotonic medium. The corresponding regulatory behaviour is characterized by rapid cell swelling within seconds due to the high water permeability of cell membranes (29), followed by a shrinkage of the cell in the time range of up to an hour. This regulatory volume decrease has been characterized in many epithelial and non-epithelial cell types. It has been shown that electrogenic and electroneutral ion transport mechanisms are involved. A direct influence of the corresponding hydraulic force on the ion flux can be neglected in comparison with the effect

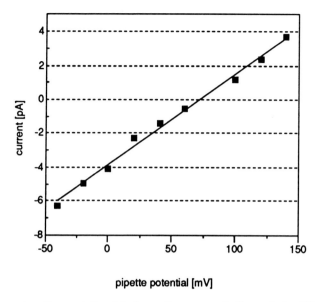

Figure 3. Current–voltage relationship for unitary currents through the SA-channel of the stretch-receptor neuron. Cell-attached patch with normal saline in the pipette. The straight line is a least-squares fit with a slope of 54 pS. Potentials are relative to the cell's membrane resting potential.

of the electric force on the transported ions (Sachs 1988, ref. 4). It is a generally accepted model that the original cell volume is adjusted by extrusion of equal amounts of potassium and chloride ions [for review see Kregenow (30); Spring and Ericson (31); Hoffmann (32); Eveloff and Warnock (33); Hoffmann and Simonsen (34)].

5.2.1 Channel-activation by hypotonic shock and suction

In opossum kidney (OK)-cells two SA-channel populations were observed. *Figure 4* shows current records of the SA-channel populations in OK-cells at increasing suction. These SA-channels differ in their single channel conductance, pressure sensitivity and pressure on half maximal activation (*Table 1*).

Recently we have found evidence suggesting that the same type of ion channel can be activated by osmotic as well as by mechanical stress:

(a) We found a one-to-one correlation in the appearance of ion channels which were activated by hypotonic shock or by negative hydrostatic pressure. Either no channel activity could be evoked in the membrane patch or in both cases the same number and size of single-channel current steps was observed (*Figure 5*).

(b) For different pipette solutions the single-channel conductance of the

Identifying and characterizing stretch-activated ion channels

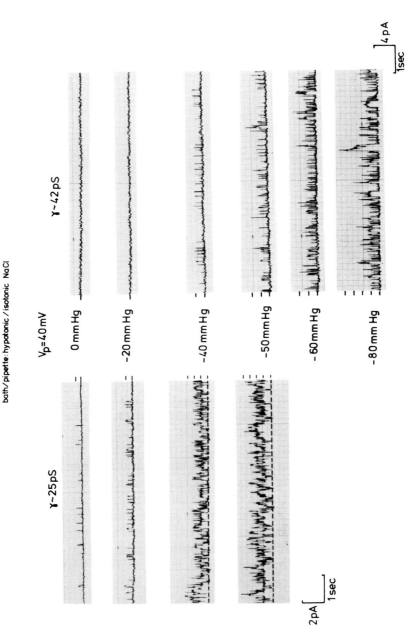

Figure 4. Pressure-dependence of ion channel activity in OK-cell membranes. Channel fluctuations of two populations of stretch-activated ion channels at different values of negative pressure applied to the pipette (interior) as indicated. The

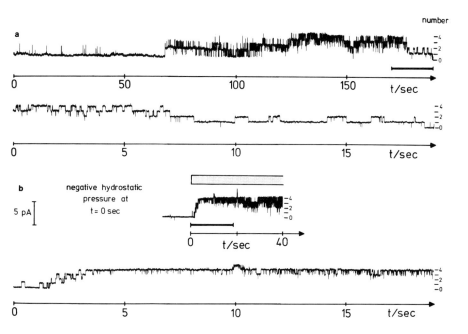

Figure 5. Activation of a channel population in an on-cell membrane patch by two different activation mechanisms. (a) 25 sec prior to time zero, the isotonic (310 mOsm NaCl) bath medium was replaced by a hypotonic (190 mOsm NaCl medium. The current trace marked by a solid bar is shown on an expanded time-scale below indicating a transient closure of the osmotically activated channels. (b) After the transient closure of the osmotically activated channels a negative pressure of 25 mmHg is constantly applied to the silent membrane patch, as indicated by the dashed bar. The pipette solution was 310 mOsm NaCl and the pipette potential 0 mV. The number of simultaneously open channels is given on the right-hand scale. Upward deflections correspond to inward currents.

osmotically evoked ion channel, measured immediately after the transient closure of the prior mechanically activated ion channel, agrees within the given limits with the values determined at a negative hydrostatic pressure (10) (*Table 2*).

5.2.2 Ion selectivity of SA-channels and their function in volume regulation

In animal cells, the SA-channels described so far are all cation-selective. In snail heart, mammalian kidney, and intestinal epithelial cells they appear to be potassium-selective. Yet for most SA-channels the reversal potential is close to zero and therefore they appear to be non-selective between different cations. (*Tables 1* and *2*). The single-channel current mostly exhibits an inward rectification. Both properties contradict an involvement of SA-

Table 2. Single-channel conductance for ion channels evoked in the on-cell configuration of OK-cells, first by application of a negative pressure of 20 mmHg to the pipette interior; second, after release of the pressure and the mechanically activated ion channel had successively closed, an osmotic stress using a hypotonic bath (190 mOsm NaCl) was applied. Data are given as mean ± SEM; n denotes the number of independently performed experiments.

Pipette solution	Bath solution Isotonic NaCl/mM Ca pressure: −25 mmHg	Hypotonic NaCl/mM Ca
Isotonic NaCl	29.9 ± 2.9 pS $n = 6$ 39.8 ± 2.0 pS $n = 6$	23.1 ± 2.2 pS $n = 11$
Isotonic KCl	22.2 ± 5.1 pS $n = 6$	17.9 ± 5.5 pS $n = 3$
75 mM CaCl$_2$	16.6 ± 0.6 pS $n = 6$	17.6 ± 0.8 pS $n = 3$
90 mM BaCl$_2$	16.9 ± 2.0 pS $n = 2$	17.9 ± 4.5 pS $n = 3$
Isotonic N-methyl–glucamine	10.0 pS $n = 1$	–
Isotonic tri-methyl ammonium–HCl	15.4 ± 0.7 pS $n = 4$	15.5 ± 0.7 pS $n = 2$
Isotonic choline– chloride	11.2 ± 0.8 pS $n = 3$	10.5 ± 1.1 pS $n = 8$
Isotonic sodium– gluconate	20.0 pS $n = 1$	19.6 ± 0.5 pS $n = 2$

channels as pathway for the proposed potassium- and chloride-efflux as response to cell swelling. Therefore, it has been proposed that a function of SA-channels is correlated with the observed high Ca-permeability. It could be demonstrated that SA-channels in leaky epithelia of choroid plexus (35), vascular endothelial cells (36), in stretch-receptors of crayfish (*Table 1*) and in opossum kidney cells (*Tables 1* and *2*) are permeable for Ca^{2+}. A corresponding increase of free cytoplasmic Ca^{2+} could activate Ca^{2+} operated K$^+$- and/or Cl$^-$-channels. For OK-cells it could be demonstrated that Ca^{2+}-dependent K$^+$-channels and a corresponding K$^+$-efflux become activated after cell exposure to a hypotonic shock (11). In parallel, a significant depolarization of OK-cells was observed. Such a cell depolarization would potentiate a potassium- and chloride-efflux. The main problem in correlating the activation of SA-channels with volume regulatory processes is the proof that SA-channels are directly involved in volume regulation and are not activated as a second-order effect due to the volume change of the cell. A further problem concerns the correlation of the time-course of SA-channel activation and cell volume change. SA-channels become activated within milliseconds after application of negative hydrostatic pressure to the pipette. The activation by a hypotonic shock is generally delayed up to two to three minutes, whereas the maximal cell volume change is reached within a few seconds. This time delay could be caused by the specific elastic properties of the cytoskeleton. The latter has been discussed as the mediator of SA-channel activation. We shall go on to discuss the proposed involvement of the cytoskeleton in volume regulation.

5.2.3 Involvement of cytoskeleton in SA-channel activation

The restoring forces of the cytoskeleton during a cell volume change, e.g. an increase, are unknown. Therefore, the mechanisms correlating a volume increase due to an obvious unfolding of the microvilli and/or microtubuli to a change of the network of the cytoskeleton is not understood (37). Furthermore, one can only speculate about the transduction of the mechanical stress to the membrane patch. There are no morphological data available on a relation between volume and surface area of eukaryotic cells during osmotic swelling. In particular, it is not known whether an increase in cell volume actually increases membrane tension. In addition it is unavoidable that the membrane patch becomes pre-conditioned by the suction procedure. Cooper *et al.* (27) have shown that repetition of determination of P_o versus p shifts the curves to lower pressure values. Therefore, the observed SA-channel activation does not necessarily reflect the behaviour of the native membrane.

Which are the observations supporting the view that the cytoskeleton is involved in SA-channel activation? Volume regulation is obviously dependent on an intact cytoskeleton (34). Pre-treatment of cells by cytochalasin B inhibits volume regulatory decrease (37) and also the correlated change of ionic permeabilities of the cell membrane (12). On the other hand, cytochalasin B increases the pressure-sensitivity of SA-channels significantly (*Table 1*). Therefore, a potentiating effect for the contribution of SA-channels in volume regulatory processes should be expected, and this was not found. It is debatable whether the cytoskeleton is directly involved in SA-channel activation. Recently, it was shown for ligand-dependent activation of membrane receptors that rearrangement of the cytoskeleton can cause a change of the enzymatic activity of membrane associated proteins like protein kinase C (for review see Niggli and Burger, ref. 38). The time-scale for SA-channel activation seems to exclude a non-mechanical type of gating mechanism. But the activity of active SA-channels is modulated by unknown mechanisms. It has been observed in chick muscle (1), in OK-cells, and in stretch-receptors, that transition from the cell-attached to the excised patch configuration significantly reduces the SA-channel open probability. In OK-cells neither the free Ca^{2+} concentration on the cytoplasmic channel side nor ATP or cAMP had any effect on the SA-channel activity under this membrane configuration. As long as specific blockers are not known for SA-channels their direct contribution to volume regulation cannot be proven.

Acknowledgement

The work was supported by the Sonderforschungsbereich 156 of the Deutsche Forschungsgemeinschaft.

References

1. Guharay, F. and Sachs, F. (1984). *Journal of Physiology*, **352**, 685.
2. Brehm, P., Kullberg, R., and Moody-Corbett, F. (1984) *Journal of Physiology*, **350**, 631.
3. Sachs, F. (1986). *Membrane Biochemistry*, **6**, 173.
4. Sachs, F. (1988). *CRC Critical Reviews in Biomedical Engineering*, **16**, 141.
5. Morris, C. E. and Sigurdson, W. J. (1989). *Science*, **243**, 807.
6. Stockbridge, L. L. and French, A. S. (1989). *Biophysical Journal*, **54**, 187.
7. Medina, I. R. and Bregestovski, P. D. (1988). *Proceedings of the Royal Society, London*, **B235**, 95.
8. Erxleben, C. (1989). *Journal of General Physiology*, **94**, 1071.
9. Erxleben, C. and Florey, E. (1988). *Pflügers Archiv für die Gesamte Physiologie*, **411**, R155.
10. Ubl, J., Murer, H., and Kolb, H.-A. (1988). *Journal of Membrane Biology*, **104**, 223.
11. Ubl, J., Murer, H., and Kolb, H.-A. (1988). *Pflügers Archiv für die Gesamte Physiologie*, **412**, 551.
12. Ubl, J., Murer, H., and Kolb, H.-A. (1988). *Pflügers Archiv für die Gesamte Physiologie*, **412**, R8.
13. Methfessel, C., Witzemann, V., Takashi, T., Mishina, M., Numa, S., and Sakmann, B. (1986) *Pflügers Archiv für die Gesamte Physiologie*, **407**, 577.
14. Taglietti, V. and Toselli, M. (1988). *Journal of Physiology*, **407**, 311.
15. Augustine, G. J., Charlton, M. P., and Smith, S. J. (1988). *Annual Reviews in Neuroscience*, **10**, 633.
16. Kater, S. B. et al. (1988). *Trends in Neuroscience*, **11**, 315.
17. Howard, J., Roberts, W. M. and Hudspeth, A. J. (1988). *Annual Review of Biophysics and Biophysical Chemistry*, **17**, 99.
18. French, A. S. (1988). *Annual Reviews in Entomology*, **33**, 39.
19. Ohmori, H. (1984). *Proceedings of the National Academy of Sciences of the USA*, **81**, 1888.
20. Hamill, O. P., Marty, A., Neher, E., Sakmann, B., and Sigworth, F. (1981). *Pflügers Archiv für die Gesamte Physiologie*, **391**, 85.
21. Sakmann, B. and Neher, E. (ed.), (1983) *Single-Channel Recording*. Plenum Press, New York.
22. Colquhoun, D. and Hawkes, A. G. (1983) In *Single-Channel Recording* (ed. B. Sakmann and E. Neher), p. 135. Plenum Press, New York.
23. Helfrich, W. (1973). *Zeitschrift für Naturforschung*, **28c**, 693.
24. Evans, E. A., Waugh, R., and Melnik, L. (1976). *Biophysical Journal*, **16**, 585.
25. Yashikawa, K., Fujimoto, T., Shimooka, T., Terada, H., Kumazawa, N., and Ishii, T. (1988). *Biophysical Chemistry*, **29**, 293.
26. Guharay, F. and Sachs, F. (1985). *Journal of Physiology*, **363**, 119.
27. Cooper, K. E., Tang, J. M., Rae, J. L., and Eisenberg, R. S. (1986) *Journal of Membrane Biology*, **93**, 259.
28. Martinac, B., Buechner, M., Delcour, A. H., Adler, J., and Kung, C. (1987). *Proceedings of the National Academy of Sciences of the USA*, **84**, 2297.
29. Finkelstein, A. (1987). *Water Movement through Lipid Bilayers, Pores, and Plasma Membranes—Theory and Reality*. Wiley-Interscience, New York.

30. Kregenow, F. M. (1981). *Annual Reviews in Physiology*, **43**, 493.
31. Spring, K. R. and Ericson, A. C. (1982). *Journal of Membrane Biology*, **69**, 167.
32. Hoffmann, E. K. (1986). *Biochemica et Biophysica Act*, **864**, 1.
33. Eveloff, J. L. and Warnock, D. G. (1987). *American Journal of Physiology*, **252**, F1.
34. Hoffmann, E. K. and Simonsen, L. O. (1989). *Physiological Reviews*, **69**, 315.
35. Christensen, O. (1987). *Nature*, **330**, 66.
36. Lansman, J. B., Hallam, T. J., and Rink, T. J. (1987). *Nature*, **325**, 811.
37. Gilles, R. (1987). In *Current Topics in Membranes and Transport*, Vol. 30 (ed. A. Kleinzeller, R. Gilles, and L. Bolis), p. 205.
38. Niggli, V. and Burger, M. M. (1987). *Journal of Membrane Biology*, **100**, 97.
39. Morris, C. E. (1990). *Journal of Membrane Biology*, **113**, 93.

SECOND-MESSENGER SYSTEMS

5

Identification of G-protein-mediated processes

A. C. DOLPHIN and R. H. SCOTT

1. Introduction

1.1. What is a G protein?

G proteins have the ability to bind guanosine triphosphate (GTP) and hydrolyse it to guanosine diphosphate (GDP). Many classes of protein possess this ability, including tubulin and the ribosomal elongation factor EF-Tu. However, in the context of signal transduction, G proteins also have recognition sites for receptors and for effectors, which may be enzymes or ion channels. Receptor activation by agonists enhances GTP binding to the G protein, and the activated G protein then selectively modifies the activity of the effector. Several recent reviews cover different aspects of G protein function (1, 2, 3).

1.2 G protein classification

The first G protein to be identified was G_s, which couples stimulatory receptors to adenylyl cyclase. It is a heterotrimer of α, β, and γ subunits. Subsequently, other species of heterotrimeric signal transducing G protein have been discovered. These include G_T (transducin), which couples rhodopsin to cyclic GMP phosphodiesterase in rod outer segments, and G_i, which couples receptors to inhibition of adenylyl cyclase and to activation of certain K^+ channels. G_o a neuronal G protein, has similar properties to G_i, although α_o does not inhibit adenylyl cyclase. It appears to have among its functions the ability to couple receptors to various ion channels (for reviews see refs 3, 4). The G proteins differ primarily in their α subunits (*Table 1*, and seem to associate with the same $\beta\gamma$ subunits, although two species of β have been identified, of molecular weights 35 and 36 kd. Other small molecular weight (21–24 kd) G proteins, including the gene products of the *ras* family, may also be involved in signal transduction, and appear to be similar to free α subunits, operating without a $\beta\gamma$ subunit (for review see ref. 5). It is the α subunit of the G protein which has the receptor and

Table 1. Species of G protein α subunits

	Number of subtypes	Molecular weight (kd)	Toxin Sensitivity
α_s	4	45, 52	CTX
α_i	3	40, 41	PTX
α_o	1	39	PTX
α_T	2 (rods, cones)	40	PTX, CTX
α_z (α_x) (ref. 48)	1	~	none known
p21 *ras*	several	21	none known
LMGS (ref. 31)	several	21–26	BTX-C3 (some)

Abbreviations CTX – cholera toxin; PTX – pertussin toxin;
BTX – botulinum toxin;
LMG – low-molecular weight G protein

effector binding regions and possesses a guanine nucleotide binding site and GTP-ase activity (6).

1.3 Characteristics of a G-protein-linked system

The receptor, signal-transducing G protein and effector are separate entities which must interact in the plane of the membrane. It is unlikely that this process is identical in all types of signal transduction, although mechanistic similarities exist. G proteins are not intrinsic membrane proteins, but are associated with its cytoplasmic face, being anchored by the hydrophobic $\beta\gamma$ subunits. Agonist activation of the receptor leads to a conformational change in the guanine nucleotide binding site of the G protein α subunit, increasing the off-rate of GDP. The binding of GTP is then favoured, and the affinity of the α subunit for the receptor is reduced. The α subunit then interacts with the effector, causing a conformational change, and alteration of activity of the effector (1). The lifetime of the activated α subunit is inherently limited by its GTP-ase activity. In some systems $\beta\gamma$ additionally enhances the GTP-ase activity of activated α. This and membrane anchorage by $\beta\gamma$ may not be its only roles, since $\beta\gamma$ has recently been identified as an activator of the enzyme phospholipase A_2 (7, 8).

Other uncertainties include whether the receptor, effector and G protein move and collide with each other randomly in the plane of the membrane ('collision coupling' which was suggested by many of the original kinetic studies on adenylyl cyclase; see ref. 1). The other possibility is that some G proteins are already closely associated in a complex with either receptor, effector, or both ('pre-coupled' systems). Although the dissociation of α from $\beta\gamma$ can be demonstrated in membrane preparations, it is not clear whether this is an essential prerequisite for G protein activation (1). In phototransduction, activated α_T may also move into the cytoplasm whereas in other systems it is anchored to the membrane by a fatty acyl group (9).

From the stoichiometry of G proteins it appears that $\beta\gamma$ and α are present in similar proportions. G_s is found in similar amount to the catalytic subunit of adenylyl cyclase. G_i, and in neural tissue G_o, are present at far higher concentrations than individual receptors.

A G-protein-mediated process, like other enzymatic processes, is highly temperature-dependent, and is also dependent on the GTP levels and therefore on the metabolic state of the cell.

2. Identification of a direct G-protein-mediated process

G proteins were initially investigated as factors coupling receptors to enzymes and regulating cytoplasmic second-messenger levels. The idea that they may also couple receptors directly to ion channels was therefore seen as somewhat unconventional. However, ion channels are oligomeric proteins which undergo conformational changes to allow ions to flow through a pore in the membrane, and it is entirely feasible for activated G proteins to induce such a conformational change or to inhibit responses to membrane potential changes.

The criteria required to show a direct effect of G protein activation on an ion channel or membrane-bound enzyme are similar, and include the necessity of ruling out that the change involves a diffusible second-messenger.

The initial indication that a G protein may be involved in any process is likely to be made on intact cells; for example, by showing that the process can be activated by a receptor normally found to be G protein-linked. Clearly, in intact systems, second messengers may be involved as mediators. An essential step is therefore to examine the receptor activation or inhibition of the enzyme or ion channel in a cell-free system, such as a membrane preparation. For ion channels, the use of patch-clamp techniques with either cell-attached or isolated patches has proven invaluable in this context. In a cell-attached patch (*Figure 1*), direct G-protein-coupled channels under the patch will not be influenced by agonists applied to the cell membrane outside the patch, but will be influenced by agonists present in the patch pipette (see Figure 3, in ref. 10). In contrast, when a patch is excised, such channels will be influenced by G protein activators or particular activated G proteins applied to the accessible cytoplasmic surface of an inside-out patch (11). Agonists applied to the accessible external surface of an outside-out patch should only require GTP/Mg^{2+} in the patch-pipette for their effect on the ion channel to be evident. Neither ATP nor other co-factors should be required. However, it must be remembered that isolated patches may well harbour membrane-bound or membrane-anchored enzymes and cytoskeletal elements. Depending on the configuration of the patch sucked into the pipette, a small amount of cytoplasm may also be retained. The reversibility of an agonist effect in an

Identification of G-protein-mediated processes

Second messenger mediated process

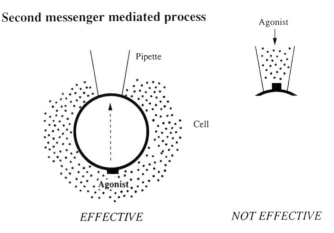

Direct or G protein mediated process

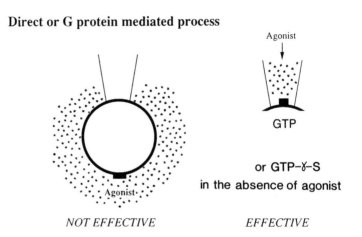

Figure 1. Schematic diagram of mechanisms which mediate modulation of ion channels by neurotransmitters and hormones. If a process is mediated by a second-messenger then ion-channel activity recorded from a cell-attached patch is modulated by an agonist applied outside the patch (■ signifies activated receptor). In contrast, in an excised inside-out patch no modulation of ion channels occurs following receptor activation by an agonist.

If a direct mechanism or a G protein mediates the process, an agonist applied to the cell but not to the patch does not result in modulation of ion channel activity. In contrast, agonist present in the patch-pipette will modulate ion channel activity recorded from an excised inside-out patch if GTP or a GTP analogue has access to the cytoplasmic face of the membrane. GTP-γ-S and other relatively non-hydrolysable analogues may modulate ion channel activity by activating G proteins after slow dissociation of GDP even in the absence of any agonist.

isolated patch should rule out the mediation of a phosphorylation–dephosphorylation process. Nevertheless, the ultimate proof of direct coupling of an ion channel to a receptor by a G protein will only come from reconstitution experiments with cloned or otherwise pure proteins.

Similar provisos apply to the assessment of G-protein-linked enzyme activity. One of the problems which may be encountered is that the retention of activity of the enzyme or ion channel in a membrane preparation, or one that has been solubilized or further purified, may be difficult. Post-translational modification such as phosphorylation, or the presence of cytoplasmic co-factors may be essential prerequisites for activation, although not necessarily taking part in the signal transduction process itself. For example, it appears that most high-threshold calcium channels require phosphorylation for activity (12). It is possibly for this reason that they are rapidly lost from isolated patches, making experiments on G protein interaction with these channels difficult to perform (11).

2.1 Summary of criteria for identification of a G-protein-coupled receptor

A G-protein-coupled receptor can be identified by several biochemical criteria.

- The affinity of the receptor for agonists is reduced by GTP or non-hydrolysable GTP analogues.
- Agonists increase the off-rate of GDP bound to the G protein.
- Purified receptor has a low affinity for agonist, and this is increased by reconstitution with the relevant G protein.

2.2 Summary of criteria for identification of a G protein-mediated process

- Activation of the receptor by agonist requires only GTP and Mg^{2+} for a response to be observed.
- The effect of agonist on the enzyme or ion channel can be mimicked irreversibly by non-hydrolysable analogues of GTP.
- The purified enzyme or ion channel can be affected by addition or purified activated G protein.

3. Experimental strategies for the identification of G-protein-mediated signal transduction

From the criteria in the previous sections it is clear that the identification of a G-protein-mediated process can be attempted from either end of the signal transduction cascade, either examining ion channel or enzyme activity, or investigating receptor-associated events. Eventually the aim must be to study the complete system.

3.1 Evidence from binding studies for receptor coupling to G proteins

When the receptor is associated with a G protein in the membrane it shows high affinity for binding agonists. The formation of this R*-G complex was found to be induced by agonist in the original studies on G_s (1), but there is evidence that a degree of pre-coupling of R and G may occur (13, 14). In the absence of GTP, the lifetime of R*-G is prolonged but when GTP is added it will rapidly displace the GDP bound to G and disrupt the R*-G complex, so that the receptor reverts to its low affinity state. The shift in affinity has been observed to be more marked for G_i than G_s-linked receptors (13, 15). This may reflect the fact that receptors such as the β-adrenergic receptor do not appear to be closely associated with G_s, and thus in membrane preparations, formation of the R*-G_s complex in the presence of agonist may be limited. No shift in affinity is generally observed for antagonists. A concentration of 100 μM GTP usually produces maximal reduction in affinity, but it must be borne in mind that GTP is highly unstable, and is also constantly hydrolysed by G proteins and other GTP-ases during the incubation process. Non-hydrolysable GTP analogues will avoid these disadvantages; for example, guanylylimido-diphosphate (GMP–PNP or GppNHp) (Sigma, 10^{-6}–10^{-4}M). Following solubilization, receptors (particularly those linked to G_i/G_o) very often retain their GTP sensitivity, suggesting that an R-G complex has been solubilized (16, 17). The identification of the G protein which remains associated to the receptor in a purified preparation of receptor is likely to prove an important means of determining which subtype of G protein couples to which receptor in particular tissues.

Now that many receptors have been cloned and sequenced, it has been observed that G-protein-coupled receptors have many features in common (for review see ref. 3). There are seven highly conserved membrane spanning regions. One particular region on the third intracellular loop appears to be required for interaction with a G protein, as shown in studies in which a chimera of the α_2 and β_2 receptor was cloned and expressed (18). This finding appears to be applicable to other G-protein-coupled receptor subtypes; marked differences are shown in this loop, suggesting that it may instill some specificity in the coupling.

3.2 Investigation of G protein GTP-ase activity

The intrinsic GTP-ase activity of isolated G proteins is low, and is limited by the low off-rate of GDP. GTP-ase activity is enhanced by association with agonist-activated receptor. Several studies have shown that even in the absence of agonist, the presence of receptor enhances GTP-ase activity to some extent. This is further evidence for pre-coupling of R and G in some systems (13).

G protein GTP-ase activity is characterized by a low K_M and is examined by determining the rate of hydrolysis of ^{32}P-GTP (19), in membrane preparations or further purified systems. Free ^{32}P$_i$ in the supernatant is separated from ^{32}P-GTP, since only the latter binds to activated charcoal.

3.3 Tools to modify G protein activity

3.3.1 Use of non-hydrolysable guanine nucleotide analogues

Two analogues of GTP are widely used, guanosine 5'-0-(3 thio)triphosphate (GTP-γ-S), and GMP-PNP. The former binds with higher affinity to the guanine nucleotide binding site on G proteins (21, 22), but has the disadvantage of being slowly hydrolysed. GMP-PNP and also guanylyl (β, γ methylene)-diphosphate are completely non-hydrolysable and once they are associated with the G protein, it will remain activated. Their dissociation rate is slow, particularly in the presence of Mg^{2+} (1). Guanosine 5'0-(2 thio) diphosphate (GDP-β-S) is a useful analogue of GDP, which may be used to block G-protein-mediated events in concentrations up to 1 mM. It binds competitively to the G protein, reducing GTP binding, and has the advantage over GDP in that it cannot be used as a substrate for the re-synthesis of GTP.

The purity of these compounds, particularly GTP-γ-S and GDP-β-S may vary between batches, and purification by HPLC may be necessary. They can be obtained from Boehringer-Mannheim and Sigma as either tetralithium or sodium salts. Concentrations of 1 μM or less GTP-γ-S are effective (22), although this depends on the concentration of other guanine nucleotides, particularly endogenous GDP and impurities in the GTP-γ-S competing for the binding site. Thus lower concentrations are effective in purified preparations. In relatively intact preparations, such as in whole-cell clamp experiments, when these compounds are included in the patch-pipette, higher concentrations are necessary. Even in this case effects have been observed with 6 μM GTP-γ-S (23). Photoactivatable derivatives of guanine nucleotide analogues have particular advantages for use in such preparations (see Section 3.5.3).

Different G proteins have different affinities for the guanine nucleotide analogues, and this may be used as an initial estimate of which G protein is involved (24). For both GTP and the analogues to be effective, Mg^{2+} is required, G_i requiring lower (μM) concentrations than G_s (mM Mg^{2+}) (1).

Radiolabelled guanine nucleotide analogues (^{3}H-GTP-γ-S or ^{35}S-GTP-γ-S and ^{3}H-GMP-PNP are available from Amersham and New England Nuclear) have been used in binding and autoradiographic studies to quantitate G proteins (16, 25), and as a tool in their purification (26). Agonist will increase the binding of the radiolabelled ligand to receptor-associated G proteins.

Another useful tool which mimics GTP analogues is the ion AlF_4^-. It is obtained by using NaF(5–10 mM) in the presence of $AlCl_3$ (5–10 μM). It appears to activate G proteins by binding to the terminal phosphate site which is exposed when GDP is bound to the G protein (27). Thus G protein activation by AlF_4^- has an advantage over the use of GTP analogues in that it does not require prior GDP dissociation.

3.3.2 Use of photaffinity labels to identify G proteins

G proteins can be identified by the use of radiolabelled guanine nucleotide analogues which have photo-affinity groups attached to them such that they bind to a G protein and can then be cross-linked to it by exposure to UV light. The irreversible tagging of G proteins in this manner has proven useful in the purification of novel G proteins (26). Since guanine nucleotide binding is stimulated by agonist-occupied receptor, an increase in association of the photoaffinity label may indicate which G protein is involved in the coupling. Several guanine nucleotide photoaffinity labels have been synthesized including 8-azido[γ-^{32}P]GTP (25) and 4-azidoanilido[γ-^{32}P]GTP (28), and the former is available commercially (ICN Radiochemicals). However, the yield of covalently-linked material obtained may not be high.

3.3.3 Use of bacterial toxins

Several bacterial toxins exert their effects by selectively ADP ribosylating certain GTP binding proteins. For example diphtheria toxin ADP ribosylates and inactivates the ribosomal elongation factor EF-Tu. More importantly for studies on signal transduction, cholera toxin, from *Vibrio cholerae*, ADP ribosylates G_s, retaining it in the GTP bound state and preventing GTP hydrolysis (for review see ref. 1). Pertussis toxin, from *Bordetella pertussis*, ADP ribosylates the G_i family and G_o in the GDP bound state, preventing receptor interaction and GTP binding (29, 30). The toxin C_3 from *Clostridium botulinum* ADP ribosylates several low molecular weight (21–24 kd) G proteins although not the *ras* gene product (31). Transducin is ADP ribosylated by both pertussis and cholera toxins (2).

These toxins consist of A and B subunits, the A subunit is a promoter and the B subunit a hexamer of non-identical subunits. They are non-covalently associated. The B oligomer is responsible for binding the toxin to the cell membrane. This occurs rapidly and is then followed by a slow internalization of the toxin, taking 30–90 min. The A subunit shows ADP ribosyl–transferase activity. In the case of pertussis toxin this activity is only manifest after reduction of a disulphide bond, which will occur in the cytoplasm (32). Thus, if pertussis toxin is used on isolated membranes, it must be pre-activated by treatment with a thiol reducing agent (32). Another problem of which to be aware, when using intact tissue, is that the binding subunit of the toxin may also be responsible for several phenomena, such as mitogenic activity, which do not depend on ADP-ribosylation (30). These responses can be mimicked by the B subunit alone.

Bacterial toxins can be used in signal transduction studies in two different ways. First, they can be used to interfere with the signal transduction process in intact cells or cell-free systems, to help determine which G protein is involved in a particular response. For example, pertussis toxin treatment of cultured cerebellar granule neurons prevents the $GABA_B$ agonist baclofen from inhibiting the release of the transmitter glutamate (*Figure 2*). Both cholera and pertussis toxins are useful in such studies. In contrast, functional changes following ADP-ribosylation by botulinum toxin C3 have yet to be identified, and the roles of its substrate G proteins are unknown. In some systems pertussis toxin does not prevent G_i/G_o activation by GTP analogues in the absence of receptor agonists, whereas in other systems this is the case. This may reflect the fact that R-G complexes may be formed to varying extents and under these conditions the binding of GTP analogues to G is enhanced even in the absence of agonist and this will be blocked by pertussis toxin.

The second use of these toxins is to identify G proteins in different tissues. In this case, membrane preparations are incubated with ^{32}P-NAD (nicotinamide adenine dinucleotide), and preactivated pertussis toxin or cholera toxin, or C3 which does not require activation by thiol reducing agents. Other co-factors include thymidine to inhibit endogenous poly-ADP-ribosyltransferase. Experimental details may be found in ref. 34. Cholera toxin preferentially ADP ribosylates receptor-associated G_s in the presence of agonist. It also requires association with an ADP-ribosylation factor (ARF) (for review see ref. 1). In contrast, pertussis toxin preferentially ADP ribosylates heterotrimeric GDP-bound G_i/G_o in the absence of agonist or GTP.

ADP-ribosylation can thus be used to identify the presence of different G proteins in tissues, following polyacrylamide gel electrophoretic separation and autoradiography, although it is not normally considered to be useful for quantitation, because of the variable NAD hydrolase activity present in different tissues. ADP-ribosylation in cell-free preparations can also be used to examine the effectiveness of prior treatment of intact tissue with toxin. Failure to label G proteins *post hoc* with ^{32}P-NAD and toxin indicates that they have already been completely ADP-ribosylated with endogenous NAD by the prior toxin treatment. A fluorescent derivative of NAD (etheno-NAD) can be used as a substrate for toxin catalysed ADP-ribosylation of G proteins (35). This may prove a useful tool for the study of changes in G protein conformation.

There may be several reasons for failure to observe an effect with cholera or pertussis toxin.

- The toxin is unable to enter intact cells because of the lack of binding sites on their surface (30, 32).

- The toxin is not active.

Identification of G-protein-mediated processes

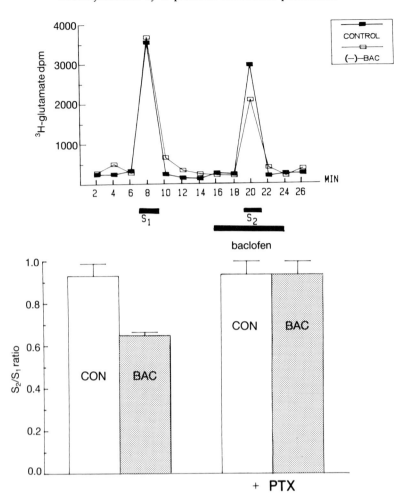

Figure 2. The release of [^{3}H]-glutamate was measured as described (33). The upper panel shows a typical experiment. Two periods of stimulated release (S_1 and S_2) were induced by 2 min incubation with 50 mM K$^+$-containing medium. (−)-baclofen (100 μM) was present before and during S_2, and reduced [^{3}H]-glutamate release in S_2 (□ compared to ■).

The lower panel shows mean results (± SEM, $n = 6$) showing a 30% reduction in the S_2/S_1 ratio due to (−)baclofen. Pre-treatment of cells for 16 h with pertussis toxin (500 ng ml^{-1} at 37°) completely prevented baclofen from inhibiting glutamate release (ref. 76).

- The G protein (or a fraction of it) may not be in the form recognized as a substrate by the toxin.
- There are no substrates for the toxin in the cell, for example, because of a species-specific mutation in the G protein.

These possibilities must be considered by performing appropriate controls before concluding that the response is not mediated by a G protein sensitive to the particular toxin. Although the toxins are effective in most mammalian tissues, this may not be the case in all species. It has been reported for example, that there are no pertussis toxin substrates in frog sympathetic neurons (36), although they have been reported in invertebrates including Aplysia and Helix.

A useful technique to unravel which G protein subtype is coupled to which receptor has been developed by Jakobs *et al.* (37). Cholera toxin will ADP-ribosylate (to a small extent) G proteins that are not normally its substrates, only when they are associated with a receptor in the presence of agonist. This trick can therefore be employed to determine whether activation of a particular receptor stimulates a particular G protein.

Pertussis toxin is available from List Biologicals (USA) and from PHLS-CAMR, Porton Down, Salisbury, Wilts, UK. It is not stable to freeze-thawing. Cholera toxin is available from Sigma. The toxins are effective when used at concentrations from 1–500 ng ml^{-1}. The lower the concentration, the longer the incubation required, up to 24 h. Once inside the cell the toxins act enzymatically. As well as their use on isolated cells, the toxins have also been applied to more intact tissue, such as organs or slices (38); however, the problem of penetration must not be overlooked. Treatment of intact tissue and membranes with the alkylating agent N-ethylmaleimide (at concentrations between 50 and 200 µM) has also been reported to produce a pertussis toxin-like inactivation of G proteins (39, 40). However, it is inevitably less specific than the toxin. Intravenous injection of toxin into animals has also been used with concentrations ranging from 0.1–1 µg/100 g body weight (41). The toxins do not normally pass the blood–brain barrier; thus to gain entry to the CNS, intraventricular or local intracerebral injections have been used effectively (42). Again, problems of lack of extensive penetration from the site of injection are likely to occur. In contrast, once inside the cell, both cholera and pertussis toxin will be subject to axonal transport. The amount of toxin injected is in the order of 1 µg, and animals are usually left at least one day before use of the tissue. Antibodies against pertussis toxin may be used to examine the degree of penetration of the toxin by immunohistochemical methods.

3.3.4 Anti-G-protein antibodies

Antibodies to G proteins have been raised by several groups (see, for example, refs 43–45; for a review see ref. 3), and are useful for several reasons:

- Antibodies bind quantitatively to their substrates and can be used for detection in a quantitative manner.
  Radioimmunoassay, immunoblotting, and immunohistochemistry at the light and EM level can be used with these tools (45, 47).

Identification of G-protein-mediated processes

- Antibodies have been raised against G proteins which are not toxin substrates, and against peptides synthesized from regions of the cDNA sequences for G proteins which have not yet been purified. This is the case for the G protein that may couple receptors to phospholipase C in some cell types (48).

- Antibodies can be raised against specific regions of G proteins thought to be involved in receptor or effector recognition or guanine nucleotide binding. Thus specific functions of G proteins can be disrupted (44).

- There are at least four pertussis toxin substrate G proteins with molecular weights in a narrow band, between 39 and 41 kd. These are more easily differentiated on gels by specific antibodies.

3.4 Reconstitution studies with exogenous G proteins

3.4.1 Use of activated G proteins and their subunits

Purified G proteins have been used in several systems to reconstitute activity; for example, where the endogenous G protein has been inactivated with pertussis toxin. G_o at a concentration of 0.4 nM is able to restore, in pertussis toxin treated tissue, the receptor-mediated inhibition of voltage-sensitive calcium channels (49, 50). Receptor-mediated activation of inwardly rectifying K channels requires 0.2–1 pM G_k (a subtype of G_i) (51, 52). It is also possible, using either G protein or α subunit pre-activated with GTP-γ-S, to activate the effector directly (53). In this case it is essential to ensure that no free GTP-γ-S is added together with the activated exogenous G protein since this could activate endogenous G proteins. Using these techniques it has been shown that both G_o and G_i can couple formyl-Met-Leu-Phe receptors to phospholipase C in HL60 cells (54), and that G_s can activate calcium channels independently of adenylyl cylclase (12).

Purified $\beta\gamma$ subunits have been used in several systems, and have been shown to terminate the effect of free α (1, 55). More unusually, they have also been found to activate cardiac K^+ channels (56). This is thought to occur indirectly by stimulation of phospholipase A_2 (8, 57). Low concentrations (200 nM) are effective at recombining with α and terminating its effect (55), but higher concentrations, up to 5 µM, have been used to activate retinal phospholipase A_2 (7).

There is some evidence that endogenous cytoplasmic GTP which is probably present at 10–50 µM can partially tonically activate G proteins (for review see ref. 3). There may thus be a pool of activated α subunits associated with effectors even in the absence of agonists. The addition of $\beta\gamma$ may thus appear to have a direct effect on these enzymes and ion channels, by recombining with activated α. $\beta\gamma$ might also enhance receptor-mediated signal transduction by increasing the heterotrimeric $\alpha\beta\gamma$ that is required for interaction with receptor.

Problems associated with the use of purified G proteins include: (a) complete purity must be ensured, (b) $\beta\gamma$ subunits are generally used in the presence of detergent; for example, CHAPS or Lubrol, to prevent aggregation, and these detergents may themselves have effects on the system (58). $\beta\gamma$ derived from transducin shows less tendency to aggregate, and can thus be used without detergent.

3.4.2 Use of cloned G proteins
Because of the possibility that contaminants are present in purified G proteins, and because cloned G protein genes inserted in *E. coli* allow the production of large amounts of specific subunits, cloned G proteins are a useful addition to the armoury (59). One potential drawback is that *E. coli* does not perform the same post-translational modifications, particularly myristoylation, which are important for these proteins to interact correctly with the membrane.

3.4.3 Use of mutant cell lines and site-directed mutagenesis
The first method used to obtain mutant G proteins was to select cell lines deficient in signal transduction. The initial cell line to be discovered was the S49 lymphoma mutant cyc$^-$, which is deficient in G_s and does not produce mRNA for G_s (60). This allowed the discovery that a separate G protein (G_i) mediates inhibition of adenylyl cyclase (61). Subsequently, α_i but not α_o was found to inhibit forskolin-stimulated adenylyl cyclase in this mutant. Another mutant (unc) produces G_s which cannot react with the receptor (62), and a third mutant H21a produces G_s which cannot interact with adenylyl cyclase (63). These mutants, and membranes derived from them, are extremely useful in reconstitution studies.

More recently the technique of site-directed mutagenesis has been used to obtain mutations at specific sites on G proteins. This has been performed extensively for the oncogene products of the *ras* gene family (for review see ref. 5). It remains unclear whether normal cellular p21 *ras* is involved directly in any of the signal transduction pathways. However, the use of mutants, producing, for example, a p21 *ras* protein which is permanently activated, have shed both light on its ability to activate phospholipase C (64), and also on the basis of its oncogenic activity in transformed cells.

3.5 Studies of G-protein-mediated processes in intact cells
3.5.1 Permeabilization
Unlike the bacterial toxins, GTP and GDP analogues are not able to enter intact cells, and thus these substances can only be used in membranes, or in electrophysiological experiments where access is available to the cytoplasmic surface of the cell. The use of these compounds to investigate physiological events such as secretion is therefore limited. Various techniques have been

Identification of G-protein-mediated processes

used to permeabilize different cell types, and to allow entry of guanine nucleotides. These include electroporation (65), which renders cells permanently leaky, and ATP^{4-} which permeabilizes mast cells but not other cell types (66). The bacterial cytolysin, streptolysin O has been used at a concentration of 0.4 IU ml^{-1} to permeabilize cells and allow the entry of guanine and other nucleotides (67). Another technique, which renders cells reversibly leaky, is hypo-osmotic shock. This has been used to introduce GTP-γ-S into fibroblasts, and demonstrate activation of phospholipase A$_2$ (68). In this case the cells reseal following a return to iso-osmotic medium. Delivery of GTP-γ-S by liposomes is also a possibility.

3.5.2 Whole-cell clamp and intracellular recording studies

Guanine nucleotide analogues (50–500 μM) diffuse steadily into cells from patch-pipettes of 2–5 MΩ resistance. Slower diffusion occurs from micropipettes and higher concentrations (20–40 mM) are necessary (42). The subunits of G proteins will also diffuse into cells from patch-pipettes (49, 50) and are effective within 10–20 min. The effect of antibodies has been studied by prior pressure injection 1 h previously to allow antibody–antigen binding to occur (69). Clearly, to be effective in such experiments, anti-G-protein antibodies must not only recognize the G protein in question, but must also have been shown to disrupt its function. This is the case for the antibody which prevents receptor-mediated inhibition of calcium currents in neuroblastoma cells (44). In this experimental design, no control data can be obtained before entry of the agents into the cell. However, when agents are present in the patch-pipette, an approximate, although unsatisfactory, control may be taken as the first response observed before diffusion has occurred. To circumvent this problem, it is possible to use the technique of two patch-pipettes, the first allowing control data to be obtained, following which the seal between the second patch-pipette and the cytoplasm is broken, allowing the compounds of interest to enter the cell. It is also possible to perfuse the patch-pipette, but both these techniques are technically difficult, particularly with small cells.

3.5.3 Intracellular photo-release of guanine nucleotide analogues

Photoactivatable ('caged') derivatives of guanine nucleotides presently in use have a nitro-phenylethyl group attached by a photo-labile ester bond to an oxygen or sulphur of the terminal phosphate or thiophosphate. They are described more fully in Chapter 10 by Jeffery Walker in *Cellular neurobiology: a practical approach*. Caged analogues of GTP itself, GTP-γ-S, GDP-β-S and GMP-PNP have been synthesized (70). Since these caged analogues are unable to activate G proteins, they can be included in the patch-pipette and diffuse into the cell, without affecting control responses as long as the experiments are performed in the dark. Photolysis is then effected, using a xenon flashlamp or laser, by light in the long UV range. The flash is filtered to exclude light below 320 nm which might damage the cell. The flashlamp used in this laboratory is supplied by Gert Rapp, Max-Planck-Institut für Medizinische Forschung, Heidelberg, Germany (71). It is relatively simple to mount the lamp as close as possible to the chamber from which recordings are made and to focus it on a spot as small as possible centred on the cell. The microscope optics can also be used to focus the light on to the cell (72) or it may be directed with a liquid light guide (70). The flash lasts about 0.5 msec, and photolysis is complete within 10–20 msec. The efficiency of photolysis depends on the intensity of the light and the quantum yield. Efficiencies of up to 10% have been achieved for guanine nucleotide derivatives in this laboratory, depending on the bond photolysed (ref. 73, and unpublished results). Efficiency is determined using a drop of caged compound (20 μl) in the position of the cell, and analysis by HPLC of the ratio of photolysis product to the parent caged compound.

Photolysis thus results in a concentration jump of pure free guanine nucleotide analogue within the cell. This allows the kinetics of the response to be examined accurately (72) and also allows concentration–response curves to be determined, since each flash will augment the concentration of GTP-γ-S in a stepwise manner (23).

Several precautions are necessary when using caged compounds, since the potentially damaging nitrosoacetophenone is also produced. It may prove necessary to react this with a thiol compound such as dithiothreitol, or reduced glutathione. Controls can be performed to rule out the involvement of nitrosoacetophenone by inclusion of caged ATP or ADP in the patch pipette, and liberating the same concentration of nitrosoacetophenone.

An example of the effect of photorelease of GTP-γ-S on the different components of the whole cell calcium channel current in cultured rat dorsal root ganglion neuron is shown in *Figure 3*.

4. Conclusion

We have discussed some of the techniques available for the examination of G proteins involved in signal transduction. There are likely to be a greater diversity of tools available in the future, particularly in terms of site-directed antibodies interfering with specific G protein functions. The technique of site-directed mutagenesis will also provide us with specific G protein mutants, disabled in particular parts of the signal transduction cycle. It is clear that specific G proteins are involved in many other transduction processes including olfactory transduction (74) and possibly electromechanical coupling, where there is evidence that a G_o-like protein may be associated with the dihydropyridine receptors in the skeletal muscle T tubules (75). Other roles for G proteins may be on the endoplasmic reticulum, in the release of intracellular Ca^{2+}, and in stimulus secretion coupling (66, 67). Much work remains to be done.

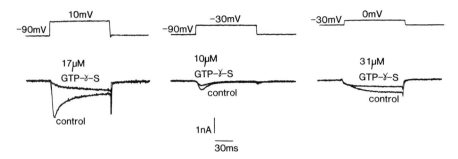

Figure 3. Whole-cell calcium channel currents recorded from cultured rat dorsal root ganglion neuron were inhibited by photo-released GTP-γ-S. The left-hand trace shows the maximum high-voltage-activated calcium channel current activated from V_H −90 mV by 100 msec voltage-step commands to V_c +10 mV. Intracellular photo-release of 17 μM GTP-γ-S preferentially inhibited the transient component of the current leaving a slowly activating, relatively non-inactivating residual current. The centre trace shows a low voltage-activated (T-type) current activated from V_H −90 mV by steps to V_c −30 mV. Intracellular photo-release of 10 μM GTP-γ-S induced significant inhibition of the current. However, lower concentrations of GTP-γ-S enhanced this current (23) suggesting that more than one G protein may be involved in its modulation. The right-hand trace shows the relatively non-inactivating high voltage-activated calcium channel current-activated from V_H−30 mV by steps to V_c 0 mV. Intracellular photo-release of 31 μM GTP-γ-S attenuated this current, but it is clearly less sensitive to GTP-γ-S than transient high and low voltage-activated currents.

All currents are illustrated following subtraction of scaled linear linkage and capacitance currents and are carried by Ba^{2+}. Photo-release of GTP-γ-S from its 'caged' precursor was achieved by light flashes (300–380 nm; 63 mJ) from a xenon flashlamp. Each flash liberated 5–7 μM GTP-γ-S.

References

1. Gilman, A. G. (1987). *Annual Reviews in Biochemistry*, **56**, 615.
2. Stryer, L. and Bourne, H. R. (1986). *Annual Reviews in Cellular Biology*, **2**, 391.
3. Milligan, G. and Houslay, M. (ed.) (1989). *G proteins* Wiley, New York.
4. Dolphin, A. C. (1990). *Annual Reviews in Physiology*, **52**, 243.
5. Dolphin, A. C. (1988) *Trends in Neuroscience*, **11**, 287.
6. Masters, S. B., Stroud, R. M., and Bourne, H. R. (1986) *Protein Engineering*, **1**, 47.
7. Jelsema, C. L. and Axelrod, J. (1987). *Proceedings of the National Academy of Sciences of the USA*, **84**, 3623.
8. Kim, D., Lewis, D. L., Graziadei, L., Neer, E. J., Bar-Sagi, D., and Clapham, D. E. (1989). *Nature*, **337**, 557.
9. Buss, J. E., Mumby, S. M., Casey, P. J., Gilman, A. G., and Sefton, B. M. (1987). *Proceedings of the National Academy of Sciences of the USA*, **84**, 7493.
10. Lipscombe, D., Kongsamut, S., and Tsien, R. W. (1989). *Nature*, **340**, 639.
11. Yatani, A., Codina, J., Reeves, J. P., Birnbaumer, L., and Brown, A. M. (1987). *Science*, **238**, 1288.
12. Chad, J. and Eckert, R. (1986). *Journal of Physiology*, **378**, 31.

13. Cerione, R. A., Codina, J., Benovic, J. L., Lefkowitz, R. J. Birnbaumer, L., and Caron, M. G. (1984). *Biochemistry*, **23**, 4519.
14. Murray, R. and Keenan, A. K. (1989). *Cellular Signalling*, **1**, 173–179.
15. Asano, T., Ui, M., and Ogasawara, N. (1985). *Journal of Biological Chemistry*, **260**, 12653.
16. Wong, Y. H., Demoliou-Mason, C. D., and Barnard, E. A. (1988). *Journal of Neurochemistry*, **51**, 114.
17. Klotz, K.-N., Keil, R., Zimmer, F. J., and Schwabe, U. (1990). *Journal of Neurochemistry*, **54**, 1988.
18. Kobilka, B. K., Kobilka, T. S., Daniel, K., Regan, J. W., Caron, M. G., and Lefkowitz, R. J. (1988). *Science*, **240**, 1310.
19. Koski, G. and Klee, W. A. (1981). *Proceedings of the National Academy of Sciences of the USA*, **78**, 4185.
20. Pfeuffer, T. and Helmreich, E. J. M. (1975). *Journal of Biological Chemistry*, **250**, 867–876.
21. Breitwieser, G. E. and Szabo, G. (1988). *Journal of General Physiology*, **91**, 469.
22. Hsia, J. A., Moss, J., Hewlett, E. L., and Vaughan, M. (1984) *Biochemical and Biophysical Research Communications*, **119**, 1068.
23. Scott, R. H., Wootton, J. F., and Dolphin, A. C. (1990). *Neuroscience*, **38**, 285.
24. Ferguson, K. M., Higashima, T., Smigel, M. D., and Gilman, A. G. (1986). *Journal of Biological Chemistry*, **261**, 7393.
25. Gehlert, D. R. and Wamsley, J. K. (1986) *European Journal of Pharmacology*, **129**, 169.
26. Waldo, G. L., Evans, T., Fraser, E. D., Northup, J. K., Martin, M. W., and Harden, T. K. (1987). *Biochemical Journal*, **246**, 431.
27. Bigay, J., Deterre, P., Pfister, C., and Chabre, M. (1985) *FEBS Letters*, **191**, 181.
28. Gordon, J. H. and Rasenick, M. M. (1988) *FEBS Letters*, **235**, 201.
29. Dolphin, A. C. (1987). *Trends in Neuroscience*, **10**, 53.
30. Ui, M., Nogimori, K., and Tamura, M. (1985). In *Pertussis Toxin* (ed. R. D. Sekura, J. Moss, and M. Vaughan). Academic Press, London.
31. Rubin, E. J., Gill, D. M., Boquet, P., and Popoff, M. R. (1988). *Molecular and Cellular Biology*, **8**, 418.
32. Sekura, R. D., Zhangi, Y.-L., and Quentin-Millet, M.-J. (1985). In *Pertussis Toxin* (ed. R. D. Sekura, J. Moss, and M. Vaughan). Academic Press, London.
33. Dolphin, A. C. and Prestwich, S. A. (1985). *Nature*, **316**, 148.
34. Katada, T. and Ui, M. (1982). *Proceedings of the National Academy of Sciences of the USA*, **79**, 3129.
35. Hingorani, V. N. and Ho, Y.-K. (1988). *Journal of Biological Chemistry*, **263**, 19804.
36. Pfaffinger, P. J. (1988). *Journal of Neuroscience*, **8**, 3343.
37. Giershik, P. and Jakobs, K.-H. (1987). *FEBS Letters*, **224**, 219.
38. Musgrave, I., Marley, P., and Majewski, H. (1987). *Naunyn-Schmiedeberg's Archiv für Pathologie*, **336**, 280.
39. Asano, T. and Ogasawara, N. (1986). *Molecular Pharmacology*, **29**, 244.
40. Fredholm, B. B., Fastbom, J., and Lindgren, E. (1986). *Acta Physiologia Scandinavica*, **127**, 381.
41. Nogimori, K., Tamura, M., Yajima, M., Ho, K., Nakamura, T., Kajikawa, N., Murayama, Y., and Ui, M. (1984). *Biochimica et Biophysica Acta*, **801**, 232.

42. Andrade, R., Malenka, R. C. and Nicoll, R. A. (1986). *Science*, **234**, 1261.
43. Giershik, P., Milligan, G., Pines, M., Goldsmith, P., Codina, J., Klee, W., and Spiegel, A. (1986). *Proceedings of the National Academy of Sciences in the USA*, **83**, 2258.
44. Milligan, G. (1988) *Biochemical Journal*, **255**, 1.
45. Huff, R. M., Axton, J. M., and Neer, E. (1985). *Journal of Biological Chemistry*, **260**, 10864.
46. Brabet, P., Dumuis, A., Sebben, M., Pantaloni, C., Bockaert, J., and Homburger, V. (1988). *Journal of Neuroscience*, **8**, 701.
47. Worley, P. F., Baraban, J. M., Van Dop, C., Neer, E. J., and Snyder, S. H. (1986). *Proceedings of the National Academy of Sciences of the USA*, **83**, 4561.
48. Fong, H. K., Yoshimoto, K. K., Eversole, C. P., and Simon, M. I. (1988). *Proceedings of the National Academy of Sciences of the USA*, **85**, 3066.
49. Hescheler, J., Rosenthal, W., Trautwein, W., and Schultz, G. (1987). *Nature*, **325**, 445.
50. Harris-Warwick, R. M., Hammond, C., Paupardin-Tritsch, D., Homburger, V., Rouot, B., Bockaert, J., and Gerschenfeld, H. M. (1988). *Neuron*, **1**, 27.
51. Yatani, A., Codina, J., Sekura, R. D., Birnbaumer, L., and Brown, A. M. (1987). *Molecular Endocrinology*, **1**, 283.
52. Kurachi, Y., Nakajima, T. and Sugimoto, T. (1986). *Pflügers Archiv für die Gesamte Physiologie*, **407**, 264.
53. Yatani, A., Codina, J., Brown, A. M., and Birnbaumer, L. (1987). *Science*, **235**, 207.
54. Kikuchi, A., Kozawa, O., Kaibuchi, K., Katada, T., Ui, M., and Takai, Y. (1986). *Journal of Biological Chemistry*, **261**, 11558.
55. Enomoto, K. and Asakawa, T. (1986). *FEBS Letters*, **202**, 63.
56. Logothetis, D. E., Kurachi, Y., Galper, J., Neer, E. J., and Clapham, D. E. (1987). *Nature*, **325**, 321.
57. Kurachi, Y., Ho, H., Sugimoto, T., Shimizu, T., Miki, I., and Ui, M. (1989). *Nature*, **337**, 555.
58. Kirsch, G. E., Yatani, A., Codina, J., Birnbaumer, L., and Brown, A. M. (1988). *American Journal of Physiology*, **254**, H1200.
59. Van Dongen, A. M. J., Codina, J., Olate, J., Mattera, Jr., Joho, R., Birnbaumer, L., and Brown, A. M. (1988). *Science*, **242**, 1433.
60. Johnson, G. L., Kaslow, H. R., Farfel, Z., and Bourne, H. R. (1980). *Advances in Cyclic Nucleotide Research*, **13**, 1.
61. Jakobs, K. H., Aktories, K., and Schultz, G. (1983). *Nature*, **303**, 177.
62. Sullivan, K. A., Miller, R. T., Masters, S. B., Beiderman, B., Heideman, W., and Bourne, H. R. (1987). *Nature*, **330**, 758.
63. Miller, R. T., Masters, S. B., Sullivan, K. A., Beiderman, B., and Bourne, H. R. (1988). *Nature*, **334**, 712.
64. Walter, M., Clark, S. G., and Levinson, A. D. (1986). *Science*, **233**, 649.
65. Knight, D. E. and Baker, P. F. (1985). *FEBS Letters*, **189**, 345.
66. Gomperts, B. D. (1983). *Nature*, **306**, 64.
67. Stutchfield, J. and Cockroft, S. (1988). *Biochemical Journal*, **250**, 375.
68. Burch, R. M., Luini, A., and Axelrod, J. (1987). *Proceedings of the National Academy of Sciences of the USA*, **83**, 7201.

69. Brown, D. A., McFadzean, I., and Milligan, G. (1989). *Journal of Physiology*, (In press.)
70. Dolphin, A. C., Wootton, J. F., Scott, R. H., and Trentham, D. R. (1988) *Pflügers Archiv für die Gesamte Physiologie*, **411**, 628.
71. Rapp, G. and Güth, K. (1988). *Pflügers Archiv für die Gesamte Physiologie*, **411**, 200.
72. Gurney, A. M., Nerbonne, J. M., and Lester, H. A. (1985). *Journal of General Physiology*, **86**, 353.
73. Dolphin, A. C., Scott, R. H., and Wootton, J. F. (1989). *Journal of Physiology*, **410**, 16P.
74. Pace, U., Hanski, E., Salomon, Y., and Lancet, D. (1985). *Nature*, **316**, 255.
75. Toutant, M., Barahanin, J., Bockaert, J., and Rouot, B. (1988) *Biochemical Journal*, **254**, 405.
76. Huston, E., Scott, R. H., and Dolphin, A. C. (1990). *Neuroscience*, **38**, 721.

6

Modifications to phosphoinositide signalling

P. JEFFREY CONN and KAREN M. WILSON

1. Introduction

Activation of a wide variety of neurotransmitter and hormone receptors results in hydrolysis of membrane phosphoinositides. It is now clear that phosphoinositide hydrolysis is the first step in a multifunctional system employed by neurotransmitter receptors for signal transduction. A number of studies over the last several years have led to a relatively detailed understanding of the nature of this response. These studies have been authoritatively reviewed by many experts in the field (1, 2). In brief, evidence suggests that activation of a phosphoinositide hydrolysis-linked receptor results in hydrolysis of all three of the major inositol-containing phospholipids by a phosphoinsitide-specific phospholipase C (PLC). These include phosphatidylinositol (PI), phosphatidylinositol-4-phosphate (PIP), and phosphatidylinositol-4,5-bisphosphate (PIP_2). It is generally thought that PIP_2 is the primary substrate for the activated PLC. Both products of PIP_2 hydrolysis, inositol-1,4,5-trisphosphate ($Ins[1,4,5]P_3$) and diacylglycerol (DAG), act as intracellular second-messengers. $Ins[1,4,5]P_3$ releases calcium from internal stores within the cell and DAG activates a phospholipid/calcium-dependent protein kinase (protein kinase C; PKC) by lowering its requirement for calcium and phospholipids. In addition, activation of phosphoinositide hydrolysis can indirectly lead to formation of other important intracellular messengers such as cyclic GMP, and arachidonic acid metabolites. The physiological responses to these second-messengers are diverse and include, among others, modulation of activity of ion channels and ion pumps, changes in neurotransmitter release, and changes in protein synthesis.

A primary mechanism for inactivation of DAG is phosphorylation to yield phosphatidic acid (PA). $Ins[1,4,5]P_3$ is metabolized by sequential dephosphorylation to inositol-1,4-bisphosphate ($Ins[1,4]P_2$), inositol-1-monophosphate ($Ins[1]P$), and finally to free inositol. Inositol and PA are then incorporated into membrane phosphoinositides and re-enter the phosphoinositide hydrolysis cycle. In addition to Ins[1]P, Ins[1,4]P_2, and Ins[1,4,5]P_3, a number of other inositol phosphates are formed with stimulation of phosphoinositide hydrolysis. Some of these compounds are direct products of hydrolysis of membrane phosphoinositides (i.e. cyclic inositol phosphates), whereas others are products of alternate routes of metabolism of Ins[1,4,5]P_3 and other hydrolysis products. Notable in the latter group are inositol-1,3,4-trisphosphate (Ins[1,3,4]P_3) and inositol 1,3,4,5-tetrakisphosphate (Ins[1,3,4,5]P_4).

In early years, agonist-induced phosphoinositide hydrolysis was generally measured by determining the effect of an agonist on incorporation of $^{32}P_i$ into PA and PI. However, this is an indirect assay that involves measurement of a secondary synthetic reaction which occurs as a result of phosphoinositide hydrolysis. In 1982, Berridge and co-workers (3) developed a sensitive method for measuring agonist-induced formation of inositol phosphates directly. These workers labelled membrane phosphoinositides with [^{3}H]inositol and measured agonist-induced formation of [^{3}H]inositol monophosphate (InsP) in the presence of lithium to inhibit inositol monophosphatase. When an agonist is added to [^{3}H]inositol-labelled cells in the presence of LiCl, radioactivity accumulates in [^{3}H]InsP over time and this greatly increases sensitivity of the assay. Not only did this assay provide a more direct measure of phosphoinositide hydrolysis, but its high sensitivity allowed measurement of relatively small increases in phosphoinositide hydrolysis that occur in response to agonists that do not have sufficient efficacy to allow measurement of $^{32}P_i$ labelling of PI and PA or formation of inositol phosphates in the absence of LiCl. In more recent years, high performance liquid chromatography (HPLC) has been employed to separate the radiolabelled inositol phosphates. This gives excellent separation of the different isomers of InsP, inositol bisphosphate (InsP$_2$) and inositol trisphosphate (InsP$_3$) and allows direct measurement of the physiologically relevant second messenger, [^{3}H]Ins[1,4,5]P_3.

An outline of a basic method that can be used for routine measurement of agonist-induced formation of InsP in the presence of LiCl in brain slices is presented in *Protocol 1*. This is a modification of the method originally described by Berridge *et al.* (3). There are a number of variations of this method, and it is likely that each investigator will modify this basic method to suit his or her particular needs. A detailed discussion of each step in this procedure and important factors to be considered when modifying each step is presented below.

Protocol 1. Measurement of agonist-induced accumulation of [^{3}H]InsP in the presence of LiCl

1. Prepare Kreb's bicarbonate buffer with 10 mM glucose (KRB) and equilibrate with O_2/CO_2 (95:5).

2. Sacrifice animals and dissect brain region of interest.
3. Slice tissue in 2 perpendicular planes (350 × 350 µM). Place slices in vial with KRB. Gas well, cap, and vortex to dissociate slices.
4. Incubate 30 min (37°C) in shaking bath.
5. During incubation, warm approximately 50 ml KRB for each vial of slices to 37°C, making sure to maintain KRB under CO_2 atmosphere while warming.
6. Wash slices with warmed KRB and remove excess buffer from gravity-packed slices.
7. Add 25 µl of gravity-packed slices to tubes containing 1 µCi each of [^{3}H]inositol in KRB.
8. Gas well, cap, and incubate for 1.5 h in shaking bath.
9. Take tubes out and allow them to sit at room temperature for 5 min before uncapping. Add LiCl to give a final concentration of 10 mM and add any antagonists desired.
10. Gas well, cap, and incubate for 15 min in shaking bath.
11. Remove tubes and allow them to sit for 5 min at room temperature before uncapping. Add agonists to reach a final volume of 300 µl.
12. Gas well, cap, and incubate for desired period of time.
13. Stop reaction by adding 3 vol. (900 µl) of chloroform/methanol (1:2) and let tubes sit for 15 min at room temperature.
14. Add 1 vol. each of chloroform and 0.5 N HCl (or water). Vortex for 1 min and separate phases by low-speed centrifugation.
15. Add 0.75 ml of the aqueous phase to columns containing Dowex-1 in the formate form.
16. Wash free [^{3}H]inositol off of columns with 15 ml of 5 mM inositol—discard.
17. Elute GPIns with 6 ml of 5 mM sodium tertraborate/60 mM ammonium formate—discard.
18. Elute InsP directly scintillation counting vials with 5 ml of 200 mM ammonium formate/0.1 M formic acid. Add 5 ml of ACS liquid counting scintillant (Amersham) to vials and determine radioactivity present in [^{3}H]InsP with a liquid scintillation counter.
19. Before reusing columns, wash inositol polyphosphates off with 15 ml of 450 mM ammonium formate/1 M formic acid followed by 15 ml of distilled water.

2. Buffer composition

The buffer most commonly used for measurement of InsP formation in brain slices is Kreb's bicarbonate buffer containing 10 mM glucose (KRB). The buffer that we use contains (in mM) NaCl, 108; KCl, 4.7; CaCl, 2.5; MgSO$_4$, 1.2; KH$_2$PO$_4$, 1.2; NaHCO$_3$, 25; and glucose, 10. The NaCl concentration of this buffer has been adjusted to allow addition of 10 mM LiCl during the procedure (*Protocol* 1, step 9). The exact composition of this buffer varies from lab to lab, and there have been no systematic studies of the effect these variations on agonist-induced phosphoinositide hydrolysis. The effect of variations of different ions on other aspects of brain slice physiology that may be relevant to phosphoinositide hydrolysis has been reviewed by Reid *et al.* (4).

Prepare buffer fresh daily. After adding bicarbonate, equilibrate the buffer by bubbling with O_2/CO_2 (95:5) to obtain a pH of 7.4. An important point to remember with bicarbonate buffer is that CO_2 can come out of solution and alter the pH of the buffer if the buffer is not continuously exposed to CO_2. To prevent CO_2 from coming out of solution (and to keep the buffer oxygenated), continuously bubble the stock buffer with O_2/CO_2 throughout the experimental period. In addition, avoid vortexing, shaking, or warming KRB unless under a CO_2 atmosphere. To achieve this, gas and cap tubes before vortexing and before each incubation. In addition, after each incubation, allow tubes to come to room temperature before uncapping and exposing to air.

Although most investigators use a bicarbonate buffer for measuring phosphoinositide hydrolysis, in preliminary experiments, we have had reasonable success using 25 mM Hepes in combination with 5 mM NaHCO$_3$. However, in general, we have obtained better results using NaHCO$_3$ as the primary buffer.

3. Preparation of slices

Sacrifice animals, rapidly remove brain and dissect the specific region that is to be studied. Slice tissue into two perpendicular planes (350 μm × 350 μm) on a McIlwain tissue chopper. Place slices in a vial containing approximately 15 ml KRB, gas, cap, and disperse slices by vortexing. After preparation of slices from the last animal, incubate the slices for 30 min in a shaking water bath at 37°C. This allows time for the tissue to recover from trauma and restore intracellular levels of ATP and other important molecules.

We routinely dissect whole cerebral cortex at room temperature. However, for regions that take longer to dissect, it is possible that dissecting on ice could increase slice viability. Since hypoxia is a major factor that can decrease cell viability in brain slices, it is important to minimize the duration of the interval between decapitation and dispersion of slices in oxygenated buffer. Thus, speed of dissection is a critical variable and should be optimized. Also, when pooling slices from a number of animals, immediately slice, gas and disperse the tissue from each animal before sacrificing the next animal. This yields significantly better results than those obtained when two or more animals are sacrificed and the tissue is pooled before slicing and dispersing. After dispersing slices, place the tissue in the incubator at 37°C while preparing slices from the next animal.

After the initial dispersion of slices, perturb the slices as little as possible. For instance, never vortex tubes containing slices. After the slices have been incubated at 37°C, they appear to become increasingly fragile with time. This may be due to the action of proteases that are active at this temperature and begin to degrade integrity of the slice.

4. Labelling slices with [^{3}H]inositol

After the 30-min incubation, pool slices from the different animals (unless different animals represent different treatment groups) and wash each vial with approximately 50 ml of warmed KRB (37°C). Cut the end off of a disposable pipettor tip (200 µl size) to give an opening with an inner diameter of approximately 2.5–3 mm. Use this tip to pipette washed slices into test-tubes containing [^{3}H]inositol in KRB. Gas with O_2/CO_2 and cap each tube, being careful to form an airtight seal in which the slices are under an O_2/CO_2 atmosphere. Polypropylene tubes (100 mm × 15 mm) and caps from Sarstedt Inc. (Princeton, NJ, USA; Rommelsdorf, Germany) are ideal for this purpose. Incubate slices with [^{3}H]inositol at 37°C.

4.1 Sources and purification of [^{3}H]inositol

[^{3}H]Inositol is available from New England Nuclear (DuPont, Wilmington, Delaware, USA; DuPont, Stevenage, Herts., UK), Amersham (Arlington Heights, Illinois, USA; Amersham, Bucks, UK) and American Radiolabelled Chemicals (ARC) (St. Louis, Missouri, USA). With time, there is spontaneous formation of an ^{3}H-labelled impurity in an aqueous solution of [^{3}H]inositol. This impurity sticks to the anion-exchange columns that are used for separation of the ^{3}H-labelled products and coelutes with inositol phosphates. Thus, it is important to maintain purity of the [^{3}H]inositol stock. Therefore, dilute the stock solution with water and store in the presence of a small amount of Dowex-1 in the formate form (see Section 8.1). Vortex the stock solution for one minute before each use and allow the Dowex to settle prior to taking an aliquot of [^{3}H]inositol.

4.2 Optimal [^{3}H]inositol and tissue concentrations

We have found that conversion of [^{3}H]inositol to [^{3}H]InsP is linear with increasing inositol concentrations between 200 nM and 1.0 µM, and that the fold increase in accumulation of radioactivity in [^{3}H]InsP induced by carbachol does not change over this concentration range. Similarly, it has been shown that incorporation of [^{3}H]inositol into phosphoinositides in brain slices is linear between 100 and 400 nM (5). Thus, we routinely add 1 μCi of [^{3}H]inositol (10–20 Ci/mmol) to tubes that will have a final volume of 300 μl (final inositol concentration = 150–350 nM).

The amount of tissue relative to the volume of buffer can be an important factor in obtaining optimal phosphoinositide hydrolysis responses. We found that the effect of serotonin on InsP accumulation is similar when 12.5–50 μl of gravity packed slices are added to tubes with a final volume of 300 μl. Variability in pipetting the slices was greater with 12.5 μl aliquots. We did not try tissue concentrations higher than 50 μl/300 μl incubation volume. However, Minneman and Johnson (6) found that norepinephrine increased [^{3}H]InsP formation linearly with increasing tissue concentrations between 25 μl and 100 μl slices per 350 μl incubation volume. The response began to decline with greater tissue concentrations. This is likely to be due to the buffering capacity of the KRB being strained by larger quantities of tissue. Janowsky *et al.* (7) found a similar biphasic response when measuring the effects of tissue concentration on carbachol-stimulated [^{3}H]InsP formation in hippocampal slices. Thus, the optimal tissue concentration should be determined for each system studied.

4.3 Duration of incubation with [^{3}H]inositol

Incorporation of radioactivity into [^{3}H]InsP increases with increasing duration of incubation with [^{3}H]inositol (*Figure 1A*). This increase is proportional in control and agonist-treated slices so that the fold increase in [^{3}H]InsP formation induced by agonists remains fairly constant with the different incubation times (*Figure 1B*). Thus, within the range studied, the duration of incubation with [^{3}H]inositol is not an important factor in terms of optimizing the response to an agonist. However, after 4 h of incubation, the slices become very fragile and such long incubations should be avoided.

Theoretically, after labelling with [^{3}H]inositol, the amount of radioactivity that appears in inositol phosphates is directly related to the mass of these sugars. However, this may not always be true if incorporation of [^{3}H]inositol into the inositol-containing lipids has not reached equilibrium. Thus, in an ideal assay, tissue slices should be incubated with [^{3}H]inositol until equilibrium is attained, and the specific radioactivity of phosphoinostides is no longer increasing with time. *Table 1* shows the time course of incorporation of [^{3}H]inositol into PI, PIP, and PIP$_2$ in cerebral cortical slices. The specific radioactivity of PI was determined at each time point but the small amounts of PIP and PIP$_2$ present prevented determination of the specific radioactivity of these lipids. However, the ratio of radioactivity in these two lipids relative to PI was relatively constant at all time points, suggesting that their specific

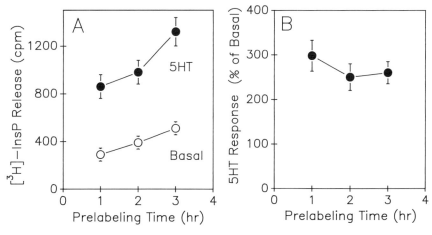

Figure 1. Effect of labelling incubation duration on serotonin-stimulated [^{3}H]IP formation. [^{3}H]InsP formation was measured in slices that were incubated with [^{3}H]inositol for various times. Labelled slices were incubated for 45 min in the presence or absence of serotonin. In panel A, data are presented as radioactivity present InsP from control and serotonin-stimulated slices. In panel B, these data are presented as a percentage of basal radioactivity present in InsP from serotonin-treated slices. Each point is the mean of four determinations. Vertical bars represent SEM.

radioactivities relative to that of PI remain constant. These data suggest that the specific radioactivity of phosphoinositides is still increasing after 4 h, suggesting that equilibrium has not been reached. Gonzalez and Crews (5) reported similar results with incubations with [^{3}H]inositol of up to 6 h. Thus, it is not practical to attempt to reach equilibrium when working with brain slices. One way to overcome this problem is by the use of cell cultures where

Table 1. Incorporation of [^{3}H]inositol into inositol-containing phospholipids

Time (h)	N	PI nCi	PI nCi/nmol	PIP nCi	PIP PI/PIP	PIP$_2$ nCi	PIP$_2$ PI/PIP$_2$
1	2	12	0.6	4.6	2.6	2.2	5.7
2	6	22 ± 2.1	1.0 ± 0.07	6.8 ± 0.7	3.2	5.0 ± 1.0	4.4
3	6	22 ± 2.3	1.7 ± 0.12	9.3 ± 1.2	3.3	5.8 ± 1.1	5.3
4	4	33 ± 1.4	1.9 ± 0.33	9.6 ± 1.2	3.4	6.1 ± 0.5	5.3

Cerebral cortical slices were incubated with [^{3}H]inositol for various times and lipids were extracted by TLC. Spots were scraped and radioactivity present in each phosphoinositide was measured. Phosphorous content in PI was measured, and specific radioactivity was determined. Data are presented as means ± SEM except for the 1-h time-point, in which each of two determinations is shown.

long incubation times with [^{3}H]inositol can be used to reach equilibrium. In primary cultures of either neurons or glia, equilibrium can be reached after 48 to 72 h incubation in the presence of [^{3}H]inositol (8).

Even in the absence of isotopic equilibrium, it is likely that an agonist-induced increase in radioactivity in [^{3}H]InsP or other inositol phosphates is indicative of an increase in InsP mass in most studies. However, in experiments in which slices from two animals that have received different *in vivo* treatments (e.g. chronic drug treatment, neuronal denervation, etc.) are compared, it is possible that the treatment could alter specific radioactivity of phosphoinositides. This could cause an artefactual change in agonist-induced phosphoinositide hydrolysis. Thus, in these experiments, it is important to control for this by determining the effect of the treatment on specific radioactivity of phosphoinositides (9) or by directly determining the effect of the *in vivo* treatment on agonist-induced changes in DAG (10, 11) or Ins[1,4,5]P_3 (12) mass (see Section 9).

4.4 Pulse-chase labelling with [^{3}H]inositol

It should be noted that a number of agonists stimulate *de novo* synthesis of phosphoinositides (13), and this may increase the rate of incorporation of [^{3}H]inositol into phosphoinositides. Such an effect could cause an artifactual increase in phospholipase C activity that is actually due to an increase in the level of substrate. This type of artefact can be controlled for by washing [^{3}H]inositol from the medium and adding unlabelled inositol prior to addition of the agonist (14). If the pulse-chase procedure is used, any inositol phosphates that are hydrolysed from newly synthesized phosphoinositides will not contain [^{3}H]inositol and will not increase the radioactive signal. For this procedure, incubate slices at a density of 200 μl gravity packed slices per millilitre of incubation medium in the presence of [^{3}H]inositol. Good results can be obtained with a 90 min incubation in the presence of 4–5 μCi [^{3}H]inositol per millilitre of incubation medium (10–20 Ci/mmol). Following this incubation, wash four times with warmed KRB containing 10 mM unlabelled inositol. Resuspend the slices in inositol-containing KRB at the original tissue concentration and incubate for an additional 60 min at 37°C. Wash again with KRB containing unlabelled inositol, aliquot gravity packed slices into test tubes, and complete the experiment as discussed above. With a 30-min incubation with unlabelled inositol, a small ligand-stimulated incorporation of [^{3}H]inositol into phosphoinositides still occurs.

5. Incubation with lithium and antagonists

After the incubation with [^{3}H]inositol, remove the tubes from the incubator and allow them to sit for 5 min at room temperature before uncapping. Allowing the incubation medium to cool before exposing to air should decrease the amount of CO_2 and O_2 that comes out of solution during the exposure to air. Uncap the tubes and add KRB containing LiCl to attain a final LiCl concentration of 5–10 mM. In addition to lithium, add antagonists and other drugs with which the slices should incubate prior to addition of the agonists. Incubate at 37°C for 15 min.

As discussed in Section 1, the purpose of adding LiCl is to inhibit inositol monophosphatase and increase sensitivity of the assay by allowing measurement of accumulation of radioactivity in [^{3}H]InsP over relatively long periods of exposure to an agonist. LiCl is not necessary if the response is to be measured on a more physiologically relevant time-scale or if the primary object is to measure formation of the inositol polyphosphates. Unfortunately, there are no cell permeable inhibitors of inositol-1,4-5-trisphosphatase that can be used to increase sensitivity of assays in which Ins[1,4,5]P$_3$ is to be measured.

6. Incubation with agonist

After the incubation with LiCl, allow samples to come to room temperature, uncap, and add agonists. We generally bring the tubes to a final volume of 300 μl. After adding appropriate drugs, gas vigorously, cap and incubate slices for the desired period of time at 37°C in a shaking water bath.

6.1 Duration of incubation and agonist

The optimal duration of incubation with agonist will depend on the particular experiment being performed. In the presence of serotonin and LiCl, levels of [^{3}H]InsP increase linearly with time for 45 min. At this point, InsP levels continue to rise but with a more shallow slope that is similar in serotonin-treated and untreated slices. Thus, there is no change in the percentage of stimulation above basal levels after 45 min. We found similar results in choroid plexus, but in this case the percentage increase was maximal at 30 min. Minneman and Johnson (6) reported that the fold increase in [^{3}H]InsP accumulation in cerebral cortical slices induced by norepinephrine continues to increase for up to 180 min. Thus, the optimal duration of incubation with agonist may be different with different agonists and brain areas.

6.2 Incubation with agonist in the absence of LiCl

Certain problems can be associated with the prolonged agonist incubations that are commonly used in assays that employ LiCl to inhibit inositol monophosphatase. For instance, when using this assay to measure the effect of an *in vivo* treatment on receptor sensitivity, remember that receptor desensitization could occur within the time course of an experiment. Thus, if chronic treatment of an animal with a drug results in receptor desensitization *in vivo*, this may not be detectable because the desensitized state of the receptor may be present in slices from both control and experimental animals after the first several minutes of agonist incubation. In addition, it may be more likely that artifactual increases in phosphoinositide hydrolysis mediated by increased *de novo* synthesis of phosphoinositides will be encountered with prolonged incubation times (see Section 3.4).

Because of the potential problems that can be associated with prolonged agonist incubations, it is often preferable to use relatively short incubation periods and measure formation of the physiologically relevant second messenger, Ins[1,4,5]P_3, directly. The optimal duration of exposure to agonist when measuring Ins[1,4,5]P_3 formation will vary depending on the system being studied, but may be on a time-scale of seconds rather than tens of minutes.

6.3 Non-receptor-mediated increases in phosphoinositide turnover

Many compounds induce an increase in inositol phosphate formation that is apparently independent of activation of a phosphoinositide hydroysis-linked receptor. This effect can be seen with a number of structurally distinct compounds and is especially prevalent at high (millimolar) concentrations of various drugs. For instance, micromolar concentrations of serotonin induce phosphoinositide hydrolysis in rat cerebral cortical slices by activating the 5HT-2 subtype of serotonin receptor (9, 15). However, at higher concentrations (1 mM and greater) serotonin induces a seemingly non-specific increase in phosphoinositide hydrolysis that is not sensitive to serotonin antagonists (15). A similar effect has been observed with high concentrations of a variety of other agents that we have used in our studies of serotonin receptors. These include serotonin receptor agonists and antagonists such as mianserin, cinanserin, and quipazine, as well as agents that act at other receptors, such as phentolamine and propranolol (16). Interestingly, although the non-specific effects of many of these compounds were pronounced in rat brain slices, they were absent rat choroid plexus (17).

The mechanism by which various agents exert these non-specific effects is unknown. However, there are a number of ways by which a non-receptor-mediated increase in InsP formation could occur, and the effects of these compounds may be mediated by more than one mechanism. For instance, cationic amphiphilic drugs increase $^{32}P_i$ labelling of phosphoinositides and other lipids by exerting effects on enzymes involved in phospholipid synthesis (18). These drugs include a number of compounds that are commonly used to study receptor function, such as propranolol, imipramine and others. In addition, some drugs that act as detergents directly activate phospholipase C (19). Regardless of the mechanism, potential non-specific effects should be controlled for by performing appropriate pharmacological experiments to determine whether a putative agonist effect is mediated by receptor activation. In cases where antagonists are used, an incubation should be included in which the effect of the antagonist in the absence of agonist is tested. This will avoid false negatives with antagonists that stimulate phosphoinositide hydrolysis. In addition, this will control for the possibility that the antagonist directly inhibits PLC.

7. Extraction of inositol phosphates

After incubation with agonists, remove tubes from the incubator and stop the reaction. There are two methods that are commonly used for stopping the reaction and extracting inositol phosphates. These are discussed below.

7.1 Chloroform/methanol extraction

For routine assays in which formation of [^{3}H]InsP in the presence of LiCl is measured, stop the reaction with 3 vol. of chloroform/methanol (2:1). Allow the tubes to sit for 15 min at room temperature to allow time for tissue extraction. Add 1 vol. each of chloroform and water and vortex for 1 min. Use low-speed centrifugation to separate the aqueous and organic phases. The upper aqueous phase contains water soluble ^{3}H-labelled compounds and can be added directly to anion-exchange columns for separation of inositol phosphates. If agonist-induced formation of inositol polyphosphates is to be determined, it is important to use an acid extraction. We have obtained poor recovery of inositol polyphosphates using simple chloroform/methanol/water extraction as have other investigators. Thus, for recovery of inositol polyphosphates, separate the phases by addition of one part each of chloroform and dilute HCl. We found that the use of 0.5 M HCl results in good extraction of inositol polyphosphates and does not alter the elution profile of inositol phosphates when separated on Dowex columns (see ref. 17). However, caution should be taken when using higher concentrations of HCl since HCl might alter the elution profile of inositol phosphates from Dowex columns. In addition to allowing measurement of inositol polyphosphates, we obtain higher counts in our InsP fractions when 0.5 M HCl is used. Thus, we routinely separate aqueous from organic phases with 0.5 M HCl and chloroform rather than water and chloroform. It should be noted however, that the inositol cyclic phosphates (e.g. inositol 1:2-cyclic phosphate) are acid labile, and if these are to be measured the extracts should not be exposed to acid extraction (see Section 6.3).

7.2 Trichloroacetic acid extraction

When inositol phosphates are separated using HPLC, we generally stop the reaction by adding ice-cold trichloroacetic acid (TCA) (final TCA concentration = 10%). This procedure can also be used when separating the inositol phosphates on Dowex columns. If the reaction is stopped with TCA, sonicate each sample for 10 sec with a probe sonicator. Centrifuge the tubes (10 000 g

for 10 min) and transfer the supernatant to a new tube. Extract the TCA from the supernatant by washing five times with water-saturated diethyl ether. The volume of each wash should be twice the volume of the supernatant. Adjust the pH of the extract to between 6 and 7 with Tris base and use this extract either for application to Dowex columns or injection into an HPLC. If needed, samples can be stored at this stage at 4°C until use. We have obtained 70–85% recovery of standards run through this extraction procedure. Some investigators terminate reactions with perchlorate (20). However, we have experienced a selective loss of InsP and $InsP_3$ in the salt precipitate when using perchlorate and prefer the TCA extraction.

7.3 Methods for improving inositol phosphate recovery

Since trace amounts of ^{3}H-labelled inositol phosphates are usually being extracted, loss of inositol phosphates during extraction can occur because of binding of these compounds to non-specific sites. For instance, when using a TCA extraction similar to that described above, Wregett *et al.* (21) reported that inositol tetrakisphosphate ($InsP_4$) recovery was significantly reduced because of absorbance of this compound to plastic surfaces and cellulose acetate filters used in their procedure. These authors found that inclusion of a phytic acid hydrolysate containing $InsP_1$–$InsP_6$ completely overcomes the loss of $InsP_4$. It is likely that inclusion of such a hydrolysate could improve recovery of inositol phosphates regardless of the extraction procedure used. A method for preparing a phytic acid hydrolysate is discussed by Wregget *et al.* (21).

Another factor to consider is that, as mentioned above, the cyclic inositol phosphates are acid labile. Thus, if there is a need to recover cyclic compounds, it is necessary to avoid acidic conditions. To accomplish this, the method of Hughes *et al.* (22) can be used. Centrifuge the cells or slices at the end of the reaction and pour off the supernatant. Add 1 vol. of boiling water. Let the samples cool to room temperature and centrifuge for 5 min at 14 000 *g*. Extract five times with 2 vol. of water-saturated ether. Apply resultant extract to HPLC or Dowex column.

8. Separation of inositol phosphates on Dowex anion-exchange columns

After extraction of inositol phosphates, apply an aliquot of the aqueous extract directly to Dowex-1 anion-exchange columns in the formate form. Inositol and inositol phosphates can be eluted from the columns as described below.

8.1 Preparation of Dowex columns

Dowex-1 anion-exchange resins are available from a number of sources. We use resin that is 4% cross-linked and has a 200–400 mesh size. Dowex-1 is normally purchased in the chloride form and must be converted to the formate form. Begin by washing the Dowex thoroughly with distilled water. After the Dowex is washed, it can be applied to columns by cutting off the end of a disposable 1-ml plastic pipette tip and pipetting 1 ml of a 50% slurry of Dowex in water into the columns. We use polystyrene columns with a plastic filter disc and 12-ml extension funnels that are purchased from Iso Lab Inc. (Ackron, Ohio, USA). In our columns (7 mm bed diameter) this gives a bed height of 14 mm. To convert to the formate form, add 20 ml of 1 M NaOH to each column. Wash with water until the pH of the eluent is less than 9. Add 4 ml of 1 M formic acid to each column. Check to make sure that pH of eluent is less than 2. If it is not repeat formate application. Wash with water until pH of the eluent is greater than 4.5. The columns are now ready for use.

8.2 Elution of inositol phosphates

Wash free [^{3}H]inositol off of the columns with 15 ml of 5 mM inositol. Inositol phosphates with increasing numbers of phosphates can then be separated by elution with increasing concentrations of ammonium formate (3, 23). Glycero-inositol phosphate (GPIns) is eluted with 5 mM sodium tetraborate/60 mM ammonium formate (solution A); InsP with 200 mM ammonium formate/0.1 M formic acid (solution B); InsP$_2$ with 450 mM ammonium formate/0.1 M formic acid (solution C); InsP$_3$ with 750 mM ammonium formate/0.1 M formic acid (solution D); and InsP$_4$ with 1 M ammonium formate/0.1 M formic acid (solution E). *Figure 2* shows an elution profile of inositol phosphates extracted from rat choroid plexus that was prelabelled with [^{3}H]inositol and incubated for 5 min in the presence of 10 µM serotonin. In this experiment InsP$_3$ and InsP$_4$ were not separated from one another and both are likely to be present in the peak eluted with solution E.

In studies in which LiCl is used to increase accumulation of InsP[1]P, the assay can be simplified by washing columns with 15 ml of 5 mM inositol and eluting total inositol phosphates into scintillation vials with 1 ml of solution E. This allows determination of radioactivity present in total inositol phosphates from stimulated and unstimulated slices. However, it is often preferable to measure radioactivity in [^{3}H]InsP selectively since the largest signal is in InsP and removal of the other compounds increases the signal-to-noise ratio. For measurement of radioactivity in [^{3}H]InsP wash free [^{3}H]inositol from the columns with 15 ml of 5 mM inositol and discard. Next, wash GPIns from the columns with 6 ml of solution A. Elute [^{3}H]InsP directly into scintillation vials with solution B. Add aqueous scintillation cocktail and determine radioactivity using scintillation counting. We elute [^{3}H]InsP with 5 ml of solution B and add 5 ml of ACS from Amersham. This forms a gel at room temperature. Counting in a gel results in a lower counting efficiency than could be obtained with a higher ratio of scintillant to water. However, as can be seen in *Figure 2*, elution with 5 ml of solution B results in recovery of almost all of the InsP peak. We find that counting the entire InsP peak results in less variability from sample to sample. This may be due to slight differences in the elution profile from each column so that elution with a smaller volume, such as 1 or 2 ml, would result in elution of different fractions of the total InsP peak from column to column.

After eluting the inositol phosphates of interest, wash the columns with 15 ml of 450 mM ammonium formate/1 M formic acid to elute remaining ^{3}H-labelled compounds. Follow this by washing the columns with 15–20 ml of water and the columns are ready for reuse. We have used columns daily for months at a time with no change in the elution profile of Ins[1]P. For instance, when approximately 12 500 c.p.m. of [^{14}C]Ins[1]P were added to fresh columns versus columns that had been in regular use for a period of months, 11 123 ± 130 c.p.m. were recovered in the InsP peak from the old columns, and 10 904 ± 114 c.p.m. were recovered in the InsP peak from the new columns. The exact period of time that columns can be reused is unknown, but we change columns about every three to four months.

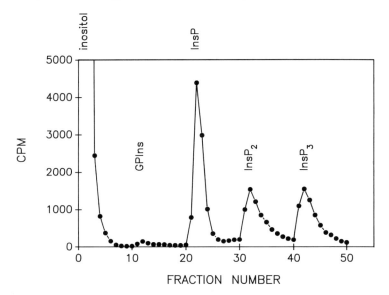

Figure 2. Separation of [^{3}H]inositol phosphates from serotonin-treated choroid plexus using Dowex-1 anion-exchange chromatography. Choroid plexus from rat was incubated with [^{3}H]inositol for 1.5 h followed by serotonin for 5 min. Inositol phosphates were extracted and applied to Dowex-1 columns as described in Sections 7.1 and 8. Radioactive compounds were then eluted with 1 ml vol. of the solutions described in Section 8.2. These include 5 mM inositol (fractions 1–10), solution A (fractions 11–20), solution B (fractions 21–30), solution C (fractions 31–40), solutions D (fractions 41–50), and solution E (fractions 51–60).

9. Separation of inositol phosphates using HPLC

When Dowex columns are used, inositol phosphates are separated on the basis of the number of phosphate groups attached to the sugar. Thus, this method cannot be used for resolution of either the isomers of $InsP_1$, $InsP_2$, and $InsP_3$ or the cyclic inositol phosphates. However, these are easily separated by using anion exchange HPLC. In addition, $InsP_4$ and inositol pentaphosphates can be identified. When inositol phosphates are to be separated using HPLC, we use the TCA extraction procedure discussed above.

9.1 HPLC equipment

Standard inositol phosphates or inositol phosphates in an extract obtained using the methods described above can be separated on a Whatman Partisal 10 SAX column (25 cm × 0.46 cm) (Whatman Inc., Clifton, NJ, USA; Whatman Ltd., Kent, England) with either an ammonium formate or ammonium phosphate elution solvent as shown in *Protocol 2*. In order to

---

Protocol 2. HPLC gradients

A. *Linear gradient (flow rate 1.25 ml/min)* (ref. 20, 26)

1. dH_2O for 6 min to remove [^{3}H]inositol
2. 26 min linear gradient dH_2O to 100% ammonium formate, pH to 3.7 with ortho-phosphoric acid to elute all $InsP_1$–$InsP_3$
3. 60 min at 100% ammonium formate to elute $InsP_4$
4. 65 min dH_2O

B. *Step gradient (flow rate 1 ml/min)* (ref. 24)

1. 0.01–0.08 M ammonium phosphate, pH 3.8 over 30 min for inositol, GPIns, inositol-1,2-cyclic phosphate, Ins[1]P, Ins[2]P, and Ins[4]P.
2. 0.2–0.28 M ammonium phosphate over 30 min for GPInsP, Ins[1,4]P$_2$, Ins[2,4]P and Ins[4,5]P$_2$.
3. 0.5–0.52 M ammonium phosphate over 30 min for GPInsP$_2$, Ins [1,4,5]P$_3$, Ins[1,3,4]P3, and Ins[2,4,5]P$_3$

C. *Elution of more polar inositol phosphates* ($InsP_4$–$InsP_6$) (ref. 27)

1. linear gradient 0.01–1.9 M ammonium formate for 30 min.
2. isocratic 1.9 M ammonium formate for 30 min.

---

lengthen the life of the column, use an injection filter (0.45 μ, ACRO LC3A, Gelman) and a guard column (Whatman). The guard column should be changed every 10–12 runs, and 100% methanol should be run through the system (1 h) when it is not going to be in use for a couple of days. Generally, we store the column in methanol over weekends. The column must then be washed with water for 1 hour prior to running the gradient. Others have also included a pre-column containing Whatman pellicular anion exchange resin (24). Different size injection loops are available, however, we have found that the 2-ml loop allowed for a practical sample size.

9.2 Elution of inositol phosphates

All solutions should be filtered (0.2 μm) and degassed prior to use. Before injection, add nucleotide standards (see below) and 100 μg of mannitol as a cold carrier to the sample. Several different gradients have been used to elute inositol phosphates on the HPLC. The most commonly used methods are presented in *Protocol 2*. We most often use the linear gradient with ammonium formate (*Protocol 2A*) because it allows for a reliable elution of $InsP_1$-$InsP_4$, the most common peaks of interest. Radioactivity in each fraction is determined in a liquid scintillation counter. The volume of the fractions collected can vary between 0.25 ml to 1.0 ml. The smaller the volume, the more defined the peaks will be, but there are limitations due to the number of samples generated as well as the capacity of the fraction collector. A typical collection pattern with the linear gradient of ammonium formate is: 13 min waste, 12 × 1 min, 35 × 0.5 min, 55 × 1 min. Using these parameters, the elution profile of inositol phosphates from neuronal cultures stimulated with norepinephrine shown in *Figure 3* was obtained. $Ins[1]P$, $Ins[1,4]P_2$, and $Ins[1,4,5]P_3$ coelute with standards run on the same elution gradient. The identity of $Ins[1,3,4]P_3$ was characterized by following the breakdown of $Ins[1,3,4,5]P_4$ in permeabilized cells. The suggested identity of the remaining isomers have been estimated from other publications (25).

Nucleotide standards (10 μM AMP, ADP, ATP, and GTP) are included in the samples as internal markers for column performance. $Ins[1]P$ coelutes with AMP, $Ins[1,4]P_2$ elutes a few fractions after ADP, and the two main $InsP_3$ isomers elute between ATP and GTP. Monitor the absorbance of nucleotides at 258 nm.

10. Measurement of $Ins[1,4,5]P_3$ and DAG mass

As discussed in Section 4.3 there are instances in which it would be preferable to use a method for examining agonist-induced increases in $Ins[1,4,5]P_3$ that does not rely on measurement of conversion of [^{3}H]inositol to ^{3}H-labelled inositol phosphates. Bredt *et al.* (12) have developed a simple and sensitive

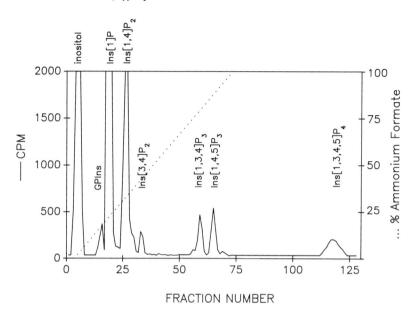

Figure 3. Separation of [³H]inositol phosphates from norepinephrine-treated neuronal cultures using HPLC. Cells were incubated with [³H]inositol for 10 days. Labelled cells were incubated for 1 h with 100 μM norepinephrine and inositol phosphates were extracted as described in Section 7.2. Inositol phosphates were eluted using the gradient described in *Protocol 2A*.

radioreceptor assay for measuring Ins[1,4,5]P_3 levels in rat brain and other tissues. To measure Ins[1,4,5]P_3 levels using this method, incubate with agonist and extract inositol phosphates using the TCA extraction procedure described in Section 7.2. Levels of Ins[1,4,5]P_3 present in these extracts can then be determined by measuring competition by tissue extracts for binding of [³H]Ins[1,4,5]P_3 to specific Ins[1,4,5]P_3 receptors in rat cerebellar membranes. This method has been described in detail by Bredt *et al.* (12).

It is also possible to measure formation of DAG, the other second-messenger that is formed with stimulation of phosphoinositide hydrolysis. The simplest and most sensitive method used for measurement of agonist-induced formation of DAG, is a radioenzymatic method developed by Preiss *et al.* (10). This method involves the measurement of conversion of DAG to [³²P]PA, when either DAG standards or tissue extracts are incubated in a buffer containing [³²P]ATP and diglyceride kinase (Lipidex Inc., Westfield, NJ, USA). The original method of Preiss *et al.* (10) does not distinguish 1-O-alkyl-diglycerides (EAG) from 1-acyl diglycerides (DAG). Only the 1-acyl diglycerides activate PKC. However, Tygai *et al.* (11) recently modified this original method so that DAG and EAG can be distinguished.

Acknowledgements

We gratefully acknowledge the valuable contributions of Dr Elaine Sanders-Bush and Dr Kenneth P. Minneman, who were collaborators in performing the studies of serotonin-stimulated phosphoinositide hydrolysis and HPLC analysis of inositol phosphates, respectively.

References

1. Abdel-Latif, A. A. (1986) *Pharmacological Reviews*, **38**, 227.
2. Fisher, S. K. and Agranoff, B. W. (1987). *Journal of Neurochemistry*, **48**, 999.
3. Berridge, M. J., Downes, C. P., and Hanley, C. P. (1982). *Biochemical Journal*, **206**, 587.
4. Reid, K. H., Edmonds, Jr., H. L., Schurr, A., Tseng, M. T., and West, C. A. (1988). *Progress in Neurobiology*, **31**, 1.
5. Gonzales, R. A. and Crews, F. T. (1984). *Journal of Neuroscience*, **4**, 3120.
6. Minneman, K. P. and Johnson, R. D. (1984). *Journal of Pharmacology and Experimental Therapeutics*, **230**, 317.
7. Janowsky, A., Labarca, R., and Paul, S. M. (1984). *Life Sciences*, **35**, 1953.
8. Gonzales, R. A., Feldstein, J. B., Crews, F. T., and Raizada, M. K. (1985). *Brain Research*, **345**, 350.
9. Conn, P. J. and Sanders-Bush, E. (1986). *Journal of Neuroscience*, **6**, 3669.
10. Preiss, J., Loomis, C. R., Bishop, W. R., Stein, R., Niedel, J. E., and Bell, R. M. (1986). *Journal of Biological Chemistry*, **261**, 8597.
11. Tygai, S. R., Burnham, D. N., and Lambeth, J. D. (1989). *Journal of Biological Chemistry*, **264**, 12977.
12. Bredt, D. S., Mourey, R. J., and Snyder, S. H. (1989). *Biochemical and Biophysical Research Communications*, **159**, 976.
13. Farese, R. V. (1983). *Endocrine Reviews*, **4**, 78.
14. Berridge, M. J., Dawson, R. M. C., Downes, C. P., Heslop, J. P., and Irvine, R. F. (1983). *Biochemical Journal*, **212**, 473.
15. Conn, P. J. and Sanders-Bush, E. (1985). *Journal of Pharmacology and Experimental Therapeutics*, **234**, 195.
16. Conn, P. J. and Sanders-Bush, E. (1987). *Journal of Pharmacology and Experimental Therapeutics*, **242**, 552.
17. Conn, P. J. and Sanders-Bush, E. (1986). *Journal of Neurochemistry*, **47**, 1754.
18. Abdel-Latif, A. A., Smith, J. P., and Akhtar, R. A. (1983). *Biochemical Pharmacology*, **32**, 3815.
19. Manning, R. and Sun, G. Y. (1983). *Journal of Neurochemistry*, **41**, 1735.
20. Ambler, S. K., Thompson, B., Solski, P. A., Brown, J. H., and Taylor, P. (1987). *Molecular Pharmacology*, **32**, 376.
21. Wreggett, K. A., Howe, L. R., Moore, J. P., and Irvine, R. F. (1987). *Biochemical Journal*, **245**, 933.
22. Hughes, A. R., Takemura, H., and Putney, J. W. (1988). *Journal of Biological Chemistry*, **263**, 10314.

23. Batty, I. R., Nahorski, S. R., and Irvine, R. F. (1985). *Biochemical Journal*, **232**, 211.
24. Dean, N. M. and Moyer, J. D. (1987). *Biochemical Journal*, **242**, 361.
25. Shears, S. B., Parry, J. B., Tang, E. K. Y., Irvine, R. F., Michell, R. H., and Kirk, C. (1987). *Biochemical Journal*, **246**, 239.
26. Irvine, R. F., Anggard, E. E., Letcher, J., and Downes, C. P. (1985). *Biochemical Journal*, **229**, 505.
27. Dean, N. M. and Moyer, J. D. (1988). *Biochemical Journal*, **250**, 493.

7

Exogenous kinases and phosphatases as probes of intracellular modulation

MICHAEL J. HUBBARD and CLAUDE B. KLEE

1. Introduction

Protein phosphorylation is a major mechanism of intracellular regulation in neural tissues, modulating a variety of enzyme activities and processes such as those involved in neurosecretion, protein and catecholamine synthesis. Since the phosphorylation state of a phosphoprotein is dynamic, determined by the relative activities of kinases and phosphatases, factors affecting these enzymes must be addressed. Experimental manipulation at the level of kinases and phosphatases is a powerful tool for investigating intracellular regulatory mechanisms involving protein phosphorylation.

Phosphoproteins, kinases, and phosphatases are abundant in most eukaryotic cells, and both kinases and phosphatases are often evolutionarily conserved. Although there are a few examples of highly specific enzyme (kinase/phosphatase) and substrate (phosphoprotein), it is clear that several broad specificity kinases and phosphatases are involved in many regulatory pathways. Ideally, experiments should be done with (endogenous) enzymes from the same cell type, and preferably the same intracellular compartment, as the phosphoregulatory system under study. However, the conserved nature and broad specificity of many kinases and phosphatases makes studies with exogenous enzymes not only feasible but often a logical place to start. Thus, one can utilize well-characterized kinases and phosphatases that are easily obtained from a major tissue source, or even supplied commercially.

An important feature of phosphorylation-mediated regulation of cellular processes is the overlap between the two major intracellular messenger systems, cyclic AMP and calcium. Calcium is now well characterized as a major second (or third) messenger in eukaryotes. Changes in the intracellular free calcium concentration are transduced to target proteins by calcium-regulated proteins such as calmodulin. The existence of calmodulin-activated protein kinases and phosphatases has introduced an exciting overlap between the calcium- and cyclic AMP-regulated systems. In this chapter, we describe general procedures used to demonstrate the role of protein kinases and phosphatases in the regulation of neuronal function, with particular emphasis on calmodulin kinase II (also called the multifunctional calmodulin-stimulated protein kinase) and the calcium/calmodulin-stimulated protein phosphatase, calcineurin. Both of these enzymes are believed to play important roles in neural tissues.

2. General approach

A biochemical approach to the study of cellular regulation by protein phosphorylation involves the following steps:

(a) establish whether the regulatory mechanism or process under study involves the phosphorylation or dephosphorylation of specific protein(s);

(b) identify the kinases that phosphorylate, and the phosphatases that dephosphorylate, the protein(s) *in vitro*;

(c) investigate *in vitro* the effects of phosphorylation and dephosphorylation on the function of the identified phosphoprotein(s);

(d) seek evidence of *in vivo* phosphorylation, and associated alterations of function, in response to physiological stimuli.

3. Detection of phosphoproteins

Whether one wants to investigate a relatively crude system, such as a subcellular fraction (e.g. synaptosomes), or to purify the individual phosphoprotein components of a modulation mechanism, an essential task is to 'freeze' the phosphorylation state of the specimen. Protein phosphatases must be inhibited to prevent dephosphorylation taking place during purification and subsequent experimental steps. It is also desirable to inhibit protein kinase activity so that spurious (artefactual) phosphorylations do not occur.

3.1 Radiolabelling with ^{32}P

This approach is widely applicable and simply involves incubating tissue, cells or subcellular fractions with [^{32}P]phosphate (intact cells) or [γ-^{32}P]ATP (broken cells). Protein kinases transfer the terminal ^{32}P group from [^{32}P]ATP to serine, threonine, and tyrosine residues of proteins, thus radiolabelling them. Labelling intact cells has the advantage that (endogenous) kinases, phosphatases and substrates are present in their native state, so any observed phosphorylations are likely to be physiologically relevant. On the other hand, the complexity of the system is increased, making it more difficult to ascribe the observed effects to any particular kinase, phosphatase or substrate. A typical radiolabelling is outlined in *Protocol 1*. Kinases utilize the Mg·ATP^{2-} adduct, not free ATP^{4-}, so magnesium should be added at low millimolar levels in broken cell experiments. Labelled phosphoproteins are readily detected by SDS–polyacrylamide gel electrophoresis followed by autoradiography. This procedure is advantageous because kinase and phosphatase activities are rapidly destroyed during denaturation of samples prior to electrophoresis, and ^{32}P-labelled proteins are simply separated from free radiolabel, either before or during electrophoresis. Moreover, transient phosphoryl intermediates (e.g. acylphosphates of ATPases) are destroyed at the alkaline pH of Laemmli gels, unlike the covalent phosphoester linkage in phosphoproteins which is resistant to a wide range of pH (pH 1–9, ≤ 37°C) (3). Autoradiographs provide a visual representation of radiolabelled proteins, separated on basis of size, from which relative abundance and extent of phosphorylation can be estimated. Two-dimensional gel electrophoresis, with isoelectric focusing in the first dimension, allows determination of the phosphoprotein's isoelectric point (4), thereby facilitating its identification. Heterogeneity (mobility shifts) during isoelectric focusing can reveal the existence of multiple phosphorylated states. Co-electrophoresis with marker proteins may lead to tentative identification of some phosphoproteins. When suitable antibodies are available, the identification can be confirmed immunologically using procedures such as immunoprecipitation or immunoblotting of the radiolabelled phosphoproteins (5–7).

Protocol 1. Labelling phosphoproteins with ^{32}P

A. *Tissues and cells*

1. Wash tissue or cells in physiological (culture) medium with reduced (0–0.2 mM) phosphate. Pre-incubation (30–60 min) with phosphate-free medium can be used to increase the intensity of subsequent radiolabelling, but this may significantly alter cellular physiology.

2. Add [^{32}P]phosphate (typically, 37–185 MBq per experiment) and incubate for an appropriate time (determine this by trial; typically 30–60 min at 37°C). Add physiological modulators (e.g. hormones) before or during this incubation period as necessary. Duplicate any additions of solvents (e.g. dimethylsulphoxide) in the controls.

3. Remove the radioactive medium carefully and rapidly wash the cells three times with physiological saline. If desired, homogenize the cells and prepare subcellular fractions (in the presence of kinase and phosphatase inhibitors); solubilize these individually, as described below.

4. Add the minimal amount of SDS sample solubilization buffer (see ref. 1), sufficient to give complete cell lysis and solubilisation (check for optical clarity). Proteinase inhibitors must be present. Sonicate the extract for 10 sec, incubate in boiling water for 3 min and then centrifuge at 14 000 g for 2–5 min to remove insoluble residues. When high levels of phospholipids are present, or excessive free radiolabel persists (e.g.

due to incomplete washing), precipitate the proteins with 10 vol. acetone at $-20°C$ and then resolubilize in SDS sample buffer.

5. Apply appropriate aliquots of the supernatant to SDS–polyacrylamide gels and subject to conventional electrophoresis, using the Laemmli discontinuous buffer system (1). Run marker proteins in parallel lanes. Stain the gels for protein (e.g. with Coomassie Blue), dry under vacuum and autoradiograph. For maximum sensitivity, perform autoradiography with pre-flashed film in an intensifying screen cassette at $-80°C$. (Alternatively, quantitate the radioactivity by direct scanning of the gels; see ref. 2.)

B. *Subcellular fractions*

1. Equilibrate the sample in a near physiological solution containing 1–5 mM magnesium (free concentration; remember to allow for chelators). Add physiological modulators as necessary.

2. Prepare a stock of 1–2 mM [^{32}P]ATP by diluting carrier free [γ-^{32}P]ATP with an appropriate amount of unlabelled ATP. Check that the pH is near neutrality.

3. Add [^{32}P]ATP to a final concentration of 0.1–0.2 mM. When NaF is used to inhibit protein phosphatases, it is preferable to add magnesium with ATP to a final concentration of 0.1–0.2 mM. Incubate for an appropriate time (determine this by trial; typically 5–10 min).

4. Solubilize and subject to SDS–polyacrylamide gel electrophoresis followed by autoradiography (see steps 4 and 5, above).

Some pitfalls of the method include:

(a) difficulties in obtaining adequate levels of labelled ATP in whole tissues and some cell types (depending on the uptake and conversion rates of [^{32}P]phosphate). Likewise, in broken cell experiments, inability to maintain sufficiently high ATP levels may reflect contaminating ATpases or a futile kinase/phosphatase cycle. ATPase inhibitors (e.g. μM vanadate) (8) or an ATP-regenerating system can be employed.

(b) failure to optimize phosphorylation conditions (increased incorporation may be offset by decreased cell viability over longer periods). Incubations should be performed under a variety of conditions (mimicking physiological stimuli appropriate to that tissue) before concluding that a protein is not phosphorylated.

(c) interference by protein phosphatases. Try adding permeant phosphatase inhibitors, such as okadaic acid (9, 10), or competing phosphatase substrates (11), to help enhance specific phosphorylation. Alternatively, [^{35}S]thiophosphate or [γ^{35}S]ATP can be used in place of ^{32}P-labelled phosphate/ATP since thiophosphorylated proteins are often more resistant to dephosphorylation (12).

3.2 Phosphate analyses

When sufficient amounts of protein are available, it may be feasible to determine the stoichiometry of covalently bound phosphate without radiolabelling. Chemical analysis of phosphate is simple but of relatively low sensitivity, thus requiring large (nanomolar) amounts of purified protein (13). Take care to avoid contamination with phospholipids (e.g. extract with ethanol:diethylether (1:1)) and nucleic acids (e.g. heat in 15% trichloroacetic acid for 15 min at 90°C). To gain increased sensitivity, the protein can be purified from ^{32}P-labelled tissue, and tested for direct incorporation of ^{32}P. In both cases, precautions should be taken to minimize changes in the phosphorylation state during purification. Use rapid freezing (e.g. freeze clamp tissues, or immerse cells in liquid nitrogen) followed by homogenization of the frozen sample in inhibitor-containing solutions to preserve the phosphorylation state close to that *in vivo*. Add phosphatase inhibitors such as fluoride (25–50 mM) and pyrophosphate (10 mM) or glycerophosphate (50–100 mM) to the solutions. Use orthovanadate (1–10 mM) to maximize detection of phosphotyrosine but be aware of the potential artefacts associated with this compound (14). Kinases are readily inhibited by inclusion of EDTA (2–5 mM) to chelate magnesium, but in most situations the rapid dilution of endogenous ATP during homogenization prevents further phosphorylation. Note that the increased ionic strength associated with these inhibitors can result in altered properties during protein purification; if necessary, compensate by decreasing the concentration of other salts.

3.3 Characterization of phosphorylation sites

Having ascertained that a protein is phosphorylated, it is advantageous to determine which residues are phosphorylated and where they lie within the primary structure. The necessary procedures can be quite straightforward if adequate amounts of purified protein are available. A detailed description is beyond the scope of this chapter, but the approach may involve: phosphoamino acid analysis; generation and mapping of peptides; and primary structural analysis of the phosphopeptides (see, for example, refs. 15 and 16).

4. Identification of protein kinases

Phosphorylation predominantly occurs at serine and threonine residues, so serine/threonine kinases are of primary interest in the effector steps of intracellular modulation. However, it is now becoming clear that tyrosine phosphorylation plays an important role in signal transduction at the plasma membrane by triggering a protein (serine/threonine) kinase cascade (17).

Many kinases have been purified and characterized on the basis of their regulation, substrate specificity and cellular distribution (18). The protein kinases reported to be present in neural tissues are listed in *Table 1*. Some of the major kinases, and their specific activators/inhibitors, are available commercially. Often, it will be obvious to test a particular kinase, based for example on its physiological or topological association with the regulation mechanism of interest. However, one should aim to obtain and test as many kinases as possible since (i) kinase substrate specificities may overlap, (ii) many phosphoproteins are phosphorylated at multiple sites, often by more than one kinase.

4.1 Assay of protein kinase activity

This involves testing for the formation of a phosphoprotein so an approach similar to that in Section 3.1 can be taken. A general kinase assay procedure is given in *Protocol 2*. Establish the assay using a known substrate for the kinase under consideration, with the following controls to ensure specificity:

- kinase omitted, or inactivated,
- magnesium omitted, or chelated,
- known substrate omitted.

Protocol 2. General procedure for protein kinase assays

1. Prepare assay mixtures, 20–50 µl final volume, in microfuge tubes. Use the following stocks: buffered solution (with composition appropriate to kinase under study); kinase (± modulators); substrate. Include controls with buffer replacing (*i*) kinase, (*ii*) substrate.
2. Bring assay mixture to temperature, usually 30°C or 37°C.
3. Start the reaction by adding [^{32}P]ATP to 0.1–0.2 mM final concentration (see *Protocol 1* for details). Mix and incubate at the above temperature for 5–10 min. Include a zero time control with trichloroacetic acid added before ATP.
4. Stop the reaction by adding 5 vol. ice-cold 12% (w/v) trichloroacetic acid and incubating on ice for at least 2 min.
5. Centrifuge the denatured proteins for 25 min at 14 000 g. Wash the pellet three times by resuspension and centrifuging from ice-cold 10% (w/v) trichloroacetic acid.
6. Add scintillant/denaturant cocktail and determine the protein-bound ^{32}P by liquid scintillation counting (typically gives > 95% efficiency). Alternatively, omit scintillant and quantitate the Cerenkov radiation ($\approx$ 60% efficiency).
7. To identify the phosphorylated species, after step 5 wash the pellet 2–3

Table 1. Protein kinases in mammalian tissues

	Subunits	$M_r \times 10^{-3}$	Composition	Substrate specificity	Inhibitors[a]	Activators[a]	Autophosphorylation	Refs
cAMP-dependent -kinase I	RI C	46 38	tetramer	broad	PKI	cAMP	+	18, 19–21
cAMP-dependent -kinase II	RII C	54 38	tetramer	broad	PKI	cAMP	+	18, 19 22
cGMP-dependent -kinase		77	dimer	broad		cGMP cAMP	+	23, 24
Myosin light -chain kinase		37–155	monomer	narrow	EGTA	Ca^{2+}	±	25–28
Calmodulin -kinase II	α β	48–55 58–60	dodecamer	broad	EGTA TFP	Ca^{2+} CaM	+	29
Calmodulin kinase I		37–42	monomer	narrow	EGTA TFP	Ca^{2+} CaM		30
Calmodulin kinase III				narrow	EGTA	Ca^{2+} CaM		31, 32
Protein kinase C		77	monomer	broad	EGTA	DG-PL Ca^{2+}, TPA	+	33, 34
Casein kinase I		37	monomer	narrow	NEM		+	18, 35 36
Casein kinase II	α β	37–44 24–28	tetramer	narrow	Heparin	Mg^{2+} polyamines	+	18, 35 36
Rhodopsin kinase		67–69	monomer	narrow				37–40

[a] The abbreviations are: PKI, protein kinase inhibitor; TFP, trifluoperazine; NEM N-ethylmaleimide; CaM, calmodulin; DG, diacylglycerol; PL, phospholipids; TPA, phorbol esters.

times with 250 μl diethylether, then solubilise and subject the sample to SDS–polyacrylamide gel electrophoresis and autoradiography (see *Protocol 1*). Note that phosphoproteins often migrate more slowly than their non-phosphorylated counterparts.

8. As an alternative to direct precipitation with trichloroacetic acid (step 4), pipet samples on to numbered segments of filter paper, then precipitate and wash them batchwise by immersions in 10% trichloroacetic acid containing 40 mM pyrophosphate, then 5% trichloroacetic acid, and finally 95% ethanol. Likewise, most proteins and peptides can be adsorbed to phosphocellulose paper, and washed with phosphoric acid (see refs 41 and 42).

With most protein substrates, the assay is stopped and the phosphoproteins are separated from the reaction mixture by precipitation with trichloroacetic acid. Alternatively (for example, for acid soluble phosphopeptides), use adsorption to charged or hydrophobic matrices (cf. *Protocol 5*) or gel permeation chromatography. Quantitate radioactivity in the isolated macromolecular fraction by liquid scintillation or Cerenkov detection. Particular kinases will require some modifications to the basic assay (to include specific cofactors or activators, for instance). The assay of calmodulin kinase II is outlined in *Protocol 3*.

To test purified proteins as potential substrates, simply substitute them in the assay for the known (control) substrate. Be aware that several complexities can arise when testing less purified fractions or tissue extracts:

- proteinase activity may destroy (or activate) the kinase, or its activators and substrates;
- ATPases may deplete ATP levels below that required for kinase activity;
- phosphatase activity may outweigh kinase activity, preventing net phosphorylation,
- the protein mixture might include kinase inhibitors; or
- other kinases.

These exigencies can all be tested for, and usually rectified following appropriate modification of the experimental conditions or allowed for with controls. When evaluating control reactions, remember that many kinases undergo autophosphorylation (*Table 1*).

If assays indicate the specific incorporation of phosphate (see controls above), it is essential to verify the presence of discrete, labelled protein species. Phosphoprotein mapping by SDS–polyacrylamide gel electrophoresis and autoradiography (Section 3.1) is usually appropriate, and may be followed by more detailed analysis of the phosphorylation sites, after appropriate scaling up (Section 3.3).

A failure to detect specific incorporation of phosphate into protein does not necessarily indicate the absence of substrates for the kinase being tested. In addition to the problems with complex mixtures noted above, consider the following possibilities:

- The substrate may already be phosphorylated, preventing incorporation of radiolabelled phosphate. Try pre-treatment with phosphatases.
- Cofactors required to provide a phosphorylatable conformation may be lacking, or species may be present which interact with the substrate thereby blocking the phosphorylation site. Try different conditions.
- Several cases are now documented of phosphorylation sites for one kinase being generated following phosphorylation by another kinase (43). Try phosphorylation with mixtures of kinases.

4.2 Primary structure of substrate

Many kinases recognize specific motifs in the primary structure of their substrates (*Table 2*, and refs 25 and 44). Accordingly, when the primary structure of a (phospho)protein is known, check to see which kinases may recognize it as a substrate. Computer programs are available to search appropriate sequences in protein databases (for example, the University of Wisconsin Genetic Computing Group's software package, or the (USA) National Biomedical Research Foundation Protein Identification Resource. This approach is more powerful when the location of the phosphorylation site in the primary structure is known (Section 3.3).

4.3 Calmodulin kinase II

Calmodulin kinase II (also known as the multifunctional calmodulin-stimulated protein kinase) has a wide tissue distribution and is abundant in brain, making it a key enzyme to test for potential involvement in neuroregulatory mechanisms. In addition to its regulation by calcium calmodulin, an important property of this enzyme is it pseudo-irreversible

---

Table 2. Typical primary structure at phosphorylation sites of cAMP-dependent protein kinase (PKA), calcium/calmodulin-stimulated protein kinase II(CaM Kinase II) and casein kinase II. Non-variant residues are indicated by conventional three-letter amino acid codes and variant residues by Www...Zzz. In some substrates threonine is phosphorylated instead of serine (Ser(P)), and conservative replacements are made at the nonvariant residues. For more details see refs. 18 and 44.

Kinase	Sequence
PKA	Xxx-Arg-Arg-Yyy-Ser(P)-Zzz
CaM kinase II	Www-Arg-Xxx-Yyy-Ser(P)-Zzz
Casein kinase II	Xxx-Yyy-Ser(P)-Glu-Asp-Glu-Zzz

activation by autophosphorylation. This characteristic, which renders calmodulin kinase II independent of calcium and colmodulin, can be used to facilitate its identification and to determine its functions (18, 26–28).

Calmodulin kinase II is highly conserved and can be readily purified from a variety of tissues (18, 25–28, 44). A method for the rapid purification of the enzyme from rat brain is described in *Protocol 3*.

Protocol 3. Purification of calmodulin kinase II.

A. *Materials*

- Buffer A: 10 mM Tris–Cl, pH 7.5 (at room temperature), 1 mM EGTA, 1 mM EDTA, 0.5 mM dithiothreitol, 1 µg/ml leupeptin, 10 µg/ml soybean trypsin inhibitor, 75 µg/ml phenylmethylsulphonyl fluoride.
- Buffer B: 40 mM Tris–Cl, pH 7.5, 0.025 mM EGTA, 50 mM NaCl, 0.1mM dithiothreitol, 2 µg/ml leupeptin, 10 µg/ml soybean trypsin inhibitor.
- Buffer C: Buffer B with 3 mM $MgCl_2$, 0.2mM $CaCl_2$. EGTA is omitted.
- Buffer D: 40 mM Tris–Cl, pH 7.5, 2 mM EGTA, 1 mM $MgCl_2$, 50 mM NaCl, 0.1 mM dithiothreitol, 2 µg/ml leupeptin, 10 µg/ml trypsin inhibitor.
- Calmodulin, purified as described (45) or obtained from a commercial source.
- Calmodulin–Sepharose (1 mg/ml) prepared as described (45) or with commercially supplied cyanogen bromide-activated Sepharose.
- Synapsin-1, purified as described (46). Other substrates can be used if synapsin-1 is not available (18, 25).

B. *Kinase assay*

1. The procedure is essentially as described in *Protocol 2*. However, since phospho-synapsin binds avidly to phosphocellulose paper (42), investigators may prefer to use this method instead of precipitation with trichloroacetic acid and centrifuging.

2. The incubation mixture contains 50 mM Tris–Cl, pH 7.5, 10 mM $MgCl_2$, 0.05 mg/ml bovine serum albumin, 0.5 mM 2-mercaptoethanol, 1 mM $CaCl_2$, 3 µM calmodulin, 0.1 mM [^{32}P]ATP (2–4 × 10^5 c.p.m./nmol), 0.1 mg/ml synapsin and up to 10^{-8} units of kinase in 50 µl.

3. The standard incubation time is 30 sec at 30°C. One unit is the amount of enzyme that catalyses the incorporation of 1 µmol P_i/min under these conditions. Specific calmodulin-dependent kinase activity is determined by subtracting control incorporation values (obtained with calcium calmodulin, and substrate omitted) from the total. The incorporation should be Ca^{2+}-dependent.

C. Purification procedure

All operations are performed at 0–4°C.

1. Homogenize 4 g of fresh rat brain in 40 ml Buffer A for 15 sec with a Polytron homogenizer. Centrifuge the homogenate for 30 min at 6000 g, wash the resulting pellet with 10 ml Buffer A, and centrifuge again as above. The combined supernatants (crude extract) contain 6–7 units of Ca^{2+}-dependent kinase activity in 46 ml.

2. To separate the kinase from endogenous calmodulin, add 1.12 g Whatman DE23 per millilitre of crude extract. Leave standing for 1 h with intermittent mixing. Filter the slurry through a course sintered glass funnel, then wash successively with (a) 20 ml Buffer A, (b) twice with 30 ml Buffer A containing 150 mM NaCl, and (c) twice with 30 ml Buffer A containing 400 mM NaCl. At least 75% of the specific calmodulin-dependent kinase activity should be recovered in the 150 mM NaCl washes (fraction 2, $\approx$ 52 ml).

3. Fractionate with ammonium sulphate (30% saturation) by adding the salt at 16.4 g per 100 ml of fraction 2, and stirring slowly for 20 min. After centrifuging for 30 min at 6000 g, dissolve the pellet in 8 ml Buffer B and clarify by centrifugation. The supernatant (fraction 3) contains 3 units of enzyme.

4. For affinity chromatography on calmodulin-Sepharose (0.6 × 4 cm), make fraction 3 0.5 mM $CaCl_2$, 3 mM $MgCl_2$ and apply it to the column (equilibrated in Buffer C) at 4 ml/h. Collect 0.5-ml fractions. Wash the column with 13 ml Buffer C, then with 20 ml Buffer C containing 200 mM NaCl. Elute the enzyme with 25 ml Buffer D. The Buffer D eluate contains $\approx$ 1.3 units of enzyme, with a specific activity of $\approx$ 3.3 units/mg protein, completely dependent on calcium and calmodulin. SDS–polyacrylamide gel electrophoresis should show 80–90% purity, with a major 55 000 M_r band.

5. Store the enzyme in aliquots at $-70°C$. After an initial 50% loss of activity, the enzyme is stable for several months.

6. Further properties of the enzyme, and alternative purification procedures, are described elsewhere (42).

---

5. Identification of protein phosphatases

Serine/threonine-specific protein phosphatases exhibit broad, overlapping substrate specificities *in vitro* and have been classified into four major types, three of which show strong structural similarities. Protein tyrosine phosphatases likewise exhibit broad substrate specificities but are structurally dissimilar to the serine/threonine-specific enzymes. Several, and perhaps all, classes of protein phosphatases are regulated (12, 47–50), thereby pointing to an active role for these enzymes in the modulation of cellular function. Accordingly, it is important to not consider phosphatases merely as tools for reversing the effects of kinases (this being their most common experimental application). Use of exogenous phosphatases to reduce the level of background (endogenous) phosphorylation might offer just as much potential as studies using exogenous kinases to boost net phosphorylation, and in some circumstances could lead to the discovery of a regulatory mechanism that would be missed in studies employing kinases only.

Like the kinases, as many types of phosphatases as possible should be tested. Relatively straightforward purification procedures have been described for all major phosphatase classes, from a variety of tissues (47–50). Unlike kinases (Section 4.2), protein phosphatases do not appear to consistently exhibit class-specific recognition of primary structural motifs in their substrates.

5.1 Assay of protein phosphatase activity

Protein phosphatase activity is most simply assayed by measuring the release of ^{32}P from radiolabelled substrate (*Protocol 4*). The assay is established with a known substrate, the general approach and controls being essentially those described for kinase assays (Section 4.1). Preparation of the substrate can be time-consuming since it requires obtaining the (dephosphorylated) substrate and an appropriate kinase, carrying out the phosphorylation, then isolating the phosophorylated substrate (freed of unincorporated radiolabel and preferably kinase) usually in its native conformation. When the substrate is an acid-insoluble phosphoprotein, the phosphatase assay can be terminated and the components separated by trichloroacetic acid precipitation. Radioactivity in the supernatant (acid soluble) fraction is quantitated and compared with the controls. An alternative approach is to use a synthetic peptide corresponding to the phosphorylation site of a characterized phosphoprotein (51). This has the advantages of reproducible and large supply of substrate at reasonable cost. Since, as synthesized, the substrate is totally unphosphorylated, the potential for incorporation of radiolabel is maximized. However, dephosphorylation of a phosphopeptide will not necessarily mimic the characteristics shown toward the parent phosphoprotein. A protocol for the preparation of a calcineurin synthetic substrate and for the assay of calcineurin with this substrate is outlined in *Protocol 5*.

Protocol 4. General procedure for protein phosphatase assays

1. Prepare ^{32}P-labelled substrate:

(a) phosphorylate the dephospho-substrate with an appropriate kinase(s) plus Mg · [γ-^{32}P]ATP (cf. *Protocol 2*, steps 1–3),

(b) stop the reaction and separate the phosphoprotein (substrate) from the kinase and unincorporated radiolabel, under non-denaturing conditions (e.g. by gel permeation chromatography or dialysis),

(c) quantitate and store the phospho-substrate.

2. Prepare assay mixtures, 20–50 μl final volume, in microfuge tubes. Use the following stocks: buffered solution (with composition appropriate for the phosphatase under study); phosphatase (± modulators).
3. Bring assay mixture to temperature, usually 30°C or 37°C.
4. Start the reaction by adding ^{32}P-labelled substrate. Mix and incubate at the above temperature for 5–10 min. Include a zero time control with trichloroacetic acid added before the substrate and controls with buffer replacing (a) phosphatase, (b) substrate.
5. Stop the reaction by adding 5 vols. ice-cold 12% (w/v) trichloroacetic acid and incubating on ice for at least 2 min.
6. Centrifuge the denatured proteins for 2–5 min at 14 000 g. Aspirate the supernatant, or a known proportion of it.
7. Add scintillant and determine the amount of released ^{32}P by liquid scintillation or Cerenkov radiation counting.
8. To identify the dephosphorylated species, follow step 7 in *Protocol 2*.
9. For acid-soluble phosphoproteins or peptides, adsorb the substrate to a solid matrix (see *Protocol 2*, step 8).

Protocol 5. Assay of calcineurin with synthetic peptide substrate

A. *Materials*

- Synthetic peptide (DLDVPIPGRFDRRVSVAAE) corresponding to the phosphorylation site on the RII subunit of cyclic AMP-dependent protein kinase (51). This is obtained commercially or synthesized by standard solid phase procedures.
- Catalytic subunit of cyclic AMP-dependent protein kinase, prepared as described (52) or obtained commercially.
- Buffer A: 40 mM Mes, pH 6.5, 0.4 mM EGTA, 0.8 mM EDTA, 4 mM $MgCl_2$, 0.1 mM $CaCl_2$, 0.1 mg/ml bovine serum albumin.

B. *Purification of the synthetic peptide by HPLC*

All operations are carried out at room temperature.

1. Pass 1 ml of the peptide solution (2 mg/ml in 50 mM ammonium bicarbonate) through a 0.45 μm filter (Millipore Co.).

2. Apply the filtered sample to a C_{18} reverse phase column (0.39 × 30 cm) equilibriated in 0.1% trifluoroacetic acid. Establish a linear 0–50% gradient of acetonitrile in 0.1% trifluoroacetic acid over 60 min, and collect 0.5 ml fractions at 1.5 ml/min. Monitor the chromatography at 225 nm (2.0 AUFS) and 280 nm (0.1 AUFS). The peptide elutes as a sharp peak, at 30 ± 2 min, and is characterised by a low 280:225 nm absorbance ratio. Confirm its identity by doing amino acid composition and/or sequence analysis.
3. Pool the appropriate peptide-containing fractions, flash evaporate, and store at −20°C. Overall yield of pure peptide should be 40–60% of the starting material. Determine the peptide concentration spectrophotometrically, using $E_{258\ nm} = 197$, or by amino acid analysis.

C. *Phosphorylation of synthetic peptide*
1. Add 45 μl of peptide solution (3.3 mM, in H_2O) and 300 μl [γ-^{32}P]ATP (1 mM, see *Protocol 1*) to 0.5 ml Buffer A. Bring the volume to 1 ml with H_2O and start the reaction with addition of 4 μg protein kinase. Incubate at 30°C for 60 min.
2. Monitor incorporation of ^{32}P into peptide by ascending thin layer chromatography of 1 μl aliquots of reaction mixture on polyethyleneimine impregnated cellulose sheets (Brinkmann Instruments Co.), using 50 mM KH_2PO_4 as solvent. The radiolabelled peptide is detected as a new radioactive spot migrating more slowly than the marker ATP. The reaction usually approaches completion at 15 min, with a final incorporation of 0.8–0.9 mol/mol.
3. Separate the labelled peptide from free radiolabel by reverse phase chromatography on a disposable C_{18} cartridge (e.g. Sep-Pak, from Waters). First prepare the cartridge by washing with 3 ml 30% acetonitrile in 0.1% trifluoroacetic acid followed by 5 ml 0.1% trifluoroacetic acid. Then apply the peptide mixture with a syringe, wash the cartridge with 0.1% trifluoroacetic acid (until < 1000 c.p.m./μl effluent) and then elute the peptide with successive 0.5 ml additions of 30% acetonitrile in 0.1% trifluoroacetic acid. Pool the radiolabelled peptide fractions (monitor with a Geiger counter), flash evaporate and store at −70°C.
4. Prior to use, dissolve the radiolabelled peptide in Buffer B and determine its concentration on the basis of radioactivity. Peptide concentration can also be measured by analytical HPLC using the extinction coefficient determined during the purification procedure.

D. *Calcineurin assay*
1. The procedure (53) is essentially as described in *Protocol 4*, except that released ^{32}P is isolated by ion-exchange chromatography. The incubation

mixture contains 1–5 µM phosphopeptide, 10^{-8}–10^{-7} M calcineurin in the presence and absence of calmodulin (equimolar with calcineurin), in a final volume of 50 µl Buffer B.

2. Terminate the reaction by adding 0.5 ml stop solution; prepare the latter by making a ten-fold dilution of 1 M potassium phosphate, pH 7, with 5% trichloroacetic acid.

3. Isolate the released inorganic phosphate (^{32}P) by chromatography on Dowex AG 50W-X8 (200–400 mesh, from BioRad) (53). Prepare 0.5-ml columns of Dowex and convert to the H$^+$ form by sequential washing with 10 ml H$_2$O, 1 ml 1M NaOH, 2 ml 1 N HCl and 4 ml H$_2$O. Apply each reaction mixture to an individual column and wash with 0.5 ml H$_2$O. Collect the combined eluates directly into scintillation vials, and quantitate the released ^{32}P.

4. One should obtain typical values of V_{max} = 1.3 µmol/min·mg, K_m = 40 µM with the synthetic peptide, at 30°C, in the presence of calmodulin. Omission of calmodulin should give about a tenfold decrease in V_{max}, and a small (about twofold) decrease in K_m.

Measuring the dephosphorylation of unknown substrates can be straightforward when the phosphoprotein is pure and singly phosphorylated. With phosphoprotein mixtures, some components of which may be multiply phosphorylated (e.g. subcellular fractions incubated with [^{32}P] ATP; see Section 3.1), the substrates for a particular phosphatase might comprise a small proportion of the total phosphoproteins present, and hence the dephosphorylation might go undetected. Accordingly, phosphoprotein and phosphopeptide mapping (Section 3) may be needed to detect site- or protein-specific dephosphorylation; the disappearance of even relatively minor phosphoprotein bands can be quite obvious upon examination of an autoradiograph.

Investigators should be wary that proteolytic degradation of phosphoproteins can also generate acid-soluble radioactivity (i.e. phosphopeptides). When assaying protein mixtures, it is necessary to confirm that the released radioactivity is free ^{32}P; for example, by checking that it is extracted into acid molybdate reagent (13).

5.2 Calcineurin

Like calmodulin kinase II, the neural isoform of calcium/calmodulin-stimulated protein phosphatase, calcineurin, is abundant in brain (particularly neurones in the dorsal root ganglia) and can therefore be expected to play important regulatory roles (49, 50). In addition to modulation by calcium and calmodulin, calcineurin activity has been shown to be affected *in vitro* by other metals, lipids and proteolysis (49, 50, 54). Moreover, myristylation of the calcium-binding (B) subunit raises the potential for regulation by targetting to membranes (49, 50).

The calmodulin-binding properties of calcineurin simplify its purification. The reader is referred to a detailed protocol for purification of the enzyme from bovine brain (55). Other approaches to purification are reviewed in ref. 49. Conflicting results were obtained in early studies of the tissue distribution and abundance of calcium/calmodulin-stimulated protein phosphatase, probably reflecting differences in the assay procedures (phosphatase activity, immunoreactivity) used. More recent studies, utilizing multiple structural and functional criteria, have confirmed that the highest levels of this enzyme exist in brain, averaging 600–1200 mg/kg (49). These studies indicate the value of using immunodetection, calcium binding (56), and calmodulin binding (57) assays as an adjunct or alternative to activity measurements (*Protocol 5*) for the detection and quantitation of calcineurin.

6. Testing for effects of phosphorylation on function

The finding that a protein is phosphorylated *in vitro* leads to the exciting prospect of being able to associate this modification with an alteration of function. Protein kinases and phosphatases can be used to probe the role of a phosphoprotein in the pathway or process under study. When the function of the phosphoprotein is already known, it might be straightforward to compare the activities of the phosphorylated and unphosphorylated forms. If a function has not been found previously, the tests should be repeated using fully phosphorylated (or dephosphorylated, as appropriate) protein; the complete modification might unmask a cryptic activity. If no function can be ascribed directly to the phosphoprotein, it may be possible (in complex systems) to detect a 'downstream' effect in the regulatory cascade. Depending on the situation, it will be preferable to either (a) prepare samples of phospho- and dephosphoprotein and test these separately in the functional assay, or (b) incorporate the phosphorylation components into the functional assay mixture, thereby making it possible to relate the extent of phosphorylation to the alteration of activity. If the second approach is adopted, control experiments must be done to verify that the expected change of phosphorylation state does actually take place in those (modified) conditions. A key feature of phosphoregulation is its reversibility, so the investigator should aim to demonstrate this *in vitro*, by using both a kinase and a phosphatase.

The lack of effect following phosphorylation may reflect a true 'silent' phosphorylation. However, it is also possible that another (unknown) activity is being modulated or, in the case of multiply phosphorylated proteins, that the overall phosphorylation is incomplete (phosphorylation by additional kinases may be required). Phosphorylation at some sites may act to modulate the dephosphorylation of other sites. Thorough characterization of phosphoproteins purified in the presence of phosphatase and kinase inhibitors (Section 3) should shed light on these possibilities.

7. Establishing a physiological role

Having identified a phosphoprotein, the (exogenous) kinases and phosphatases that recognize it, and perhaps demonstrated an association between phosphorylation state and function, the investigator may still face the hardest challenge of all: proving that these components interact and function *in vivo*.

Initially, one can question whether the results obtained using exogenous kinase and phosphatases are reproducible with endogenous (or homologous) ones. However, this experiment may be either difficult to carry out when purifying the endogenous enzymes is difficult (e.g. limited tissue supply) or unnecessary when there is reason to expect that insignificant differences exist (e.g. already documented as a highly conserved protein).

The following considerations may help to evaluate the physiological significance of the *in vitro* observations with exogenous kinases and phosphatases, and provide avenues for further investigation:

(a) Generally, physiologically relevant reactions are efficient, with large specificity constants ($k_{cat}/K_m > 10^6$ sec^{-1} M^{-1}). In other cases, a lower catalytic efficiency may be compensated for by the relatively high concentration of reactants. The rate of the phosphorylation and dephosphorylation reactions determined *in vitro* (done with near as possible physiological conditions, including enzyme and substrate concentrations) should be comparable with those observed physiologically (see ref. 58, for example).

(b) It is expected that the phosphoprotein and its kinases and phosphatases are located within the same cellular compartments. Localization studies may use subcellular fractionation and immunocytochemical approaches (preferably both); although difficult in some systems, efforts must be directed at obtaining quantitative data. Evidence is accumulating that many kinases and phosphatases are 'targeted' to specific locations, and that mechanisms exist for controlled release (translocation) of the catalytic units (31, 32, 48, 59, 60). Hence co-localization may be evident under some physiological conditions but not others.

(c) Alterations of *in vivo* phosphorylation state in response to physiological stimuli may be revealed by studies that involve detection or purification and characterization of phosphoproteins, as outlined in Section 3 (61, 62). While often strongly corroborating the *in vitro* data, the possible involvement of other kinases and phosphatases (with overlapping substrate specificities) must be kept in mind. Analyses of site-specific phosphorylation will usually be required since a key (local) phosphorylation change may oppose the net change in cellular phosphorylation. For example, although insulin promotes an overall increase in cellular phosphorylation, the activation of glycogen synthesis by insulin is associated with decreased phosphorylation at critical residues on glycogen synthase. Advanced studies of multiply phosphorylated proteins may be facilitated by the use of mass spectrometry (61).

(d) The existence of well characterized kinase and phosphatase modulators provides the investigator with powerful tools to alter the net phosphorylation state, and ascertain the effect on function. The modulators may act directly (e.g. kinase/phosphatase activators and inhibitors) or indirectly (e.g. altered level of calcium or nucleotide). G protein agonists and antagonists, adenylate cyclase and phosphodiesterase modulators, cyclic nucleotide derivatives and a specific kinase inhibitor protein have all been used to demonstrate the role of cyclic AMP-dependent protein kinase in neuronal function. Phorbol esters have played an important role in identifying the role of protein kinase C, and calcium-regulated processes have been investigated with calcium ionophores and caged calcium (31, 33, 62–70). The heat stable proteina, termed inhibitor-1 and -2, are potent, specific inhibitors of some (type 1) but not other (type 2) protein phosphatases (*Table 3*), and have been used to manipulate intracellular phosphatase activity (47). A recently identified phosphatase inhibitor, okadaic acid, is potent, specific, and membrane permeant, making it very useful for studies of *in vivo* phosphorylation (9, 10). Another approach has been to use pseudosubstrates as inhibitors. When the enzyme substrate specificity is narrow, the substrates themselves can be used as probes. The elegant studies of Selinger (11), suggesting that the lack of calcineurin underlies the light-induced degeneration of *rdgB* Drosophila mutant, is such an example.

The fundamental problem with the use of phosphorylation modulators (75) is verifying that the intended effect (usually based on *in vitro* findings) is specific to that event within the cell. For example, while potently inhibiting calmodulin-stimulated reactions *in vitro*, phenothiazines have been found to also affect other activities, making studies with these compounds open to misinterpretation (76).

(e) A related approach to that in (d) is to increase the amount of kinase or phosphatase activity directly by introducing an excess of the purified protein (73). Use of enzymes activated by proteolysis or (thio)phosphorylation might allow the requirement for physiological stimuli to be bypassed, providing strong evidence for the involvement of these enzymes in a particular cellular response. Microinjection or cell permeabilization techniques may be required to introduce some compounds in sufficient amounts; however, these procedures might significantly alter cellular physiology.

(f) The genetic approach, where feasible, remains the most efficient method to demonstrate the role of a particular enzyme or protein *in vivo*. For example, recent genetic studies have pointed to the key involvement of type 1 protein phosphatase(s) in control of the cell cycle (77). Likewise, the role of calmodulin in regulating *Paramecium* voltage-dependent potassium channels has been demonstrated in this way (78). With the rapid progress in molecular

Table 3. Protein phosphatases in mammalian tissues.

	Subunit $M_r \times 10^{-3}$	composition	Substrate specificity	Inhibitors	Activators	Refs
A. *Serine/threonine-specific phosphatases*						
Protein phosphatase 1 (PP1)						
catalytic subunit (PP1$_C$)	37	monomer	broad	I-1, I-2 okadaic, heparin, pp60^{v-src}	heparin	12, 47
glycogen form (PP1$_G$)	37 + 161	dimer	broad	PKA	glycogen, PP2A, PP2B	47 16 58
myofibril form (PP1$_M$)	37 + ?	?dimer	broad			47
inactive cytosolic form (PP1$_I$)	37 + 23	dimer			GSK-3	47
Protein phosphatase 2A (PP2A)						
catalytic subunit (PP2A$_C$)	36	monomer	broad	okadaic	heparin, polyamines, basic proteins	12, 47
native enzyme (PP2A$_1$, PP2A$_2$)	36 + (55–70)	dimer dimer/trimer	broad			
Protein phosphatase 2B (PP2B) or calcineurin	61 + 19	dimer	narrow	EGTA TFP	Ca^{2+}, calmodulin phospholipid Ni^{2+}, Mn^{2+}, Mg^{2+}	12, 49, 50 54, 71, 72
Protein phosphatase 2C (PP2C)	43	monomer	broad		Mg^{2+}	47
B. *Tyrosine-specific phosphatases*						
Placental PTPaseI1B	37	monomer	broad		Inhibitor H	73, 74
Brain PTPase 5	48	monomer	broad		Inhibitor L	12, 48

Abbreviations: PKA, phosphorylation by cAMP-dependent protein kinase; GSK-3, phosphorylation by glycogen synthase kinase 3; TFP, trifluoperazine biology techniques, the ability to create mutant kinases and phosphatases or to rescue naturally occurring mutants will become the methods of choice Molecular biology techniques might also be used to alter the expression levels of the kinases/phosphatases, their protein regulators and substrates, or to mutate the phosphorylation sites of substrates to assess the effects on function.

Acknowledgements

Michael Hubbard thanks colleagues at the Protein Phosphorylation Group (Department of Biochemistry, University of Dundee, Scotland) and the Laboratory of Biochemistry (National Cancer Institute, NIH, Bethesda, Maryland, USA) for assisting him with many of the methods and approaches described here.

References

1. Laemmli, U.K. (1970). *Nature*, **227**, 680.
2. Sullivan, D. E., Auron, P. E., Quigley, G. J., Watkins, P., Stanchfield, J. E., and Bolon, C. (1987). *Biotechniques*, **5**, 672.
3. Weller, M. (1979). *Protein Phosphorylation*. Pion. London.
4. O'Farrell, P. H. (1975). *Journal of Biological Chemistry*, **250**, 4007.
5. Kessler, S. W. (1981). In *Methods in Enzymology*, Vol. 73 (ed. J. J. Longone and H. Van Vonakis), p. 442. Academic Press, New York.
6. Gershoni, J. M. and Palade, G. E. (1983). *Analytical Biochemistry*, **131**, 1.
7. Matsudaira, P. T. (ed.) (1989). *A Practical Guide to Protein and Peptide Purification for Microsequencing*. Academic Press, New York.
8. Macara, I. G. (1980). *Trends in Biochemical Science*, **5**, 92.
9. Bialojan, C. and Takai, A. (1988). *Biochemical Journal*, **256**, 283.
10. Haystead, T. A. J., Sim, A. T. R., Carling, D., Honnor, R. C., Tsukitani, Y., Cohen, P., and Hardie, D. G. (1989). *Nature*, **337**, 78.
11. Minke, B., Rubinstein, C. T., Sahly, I., Bar-Nachum, S., Timberg, R., and Selinger, Z. (1990). *Proceedings of the National Academy of Sciences of the USA*, **87**, 113.
12. Merlevede, W. and DiSalvo, Y. (ed.) (1984–88). *Advances in Phosphatases*, Vol. 1–5. Leuven University Press, Leuven, Belgium.
13. Van Veldhoven, P. P. and Mannaerts, G. P. (1987). *Analytical Biochemistry*, **161**, 45.
14. Tracey, A. S. and Gresser, M. J. (1986). *Proceedings of the National Academy of Sciences of the USA*. **83**, 609.
15. Hubbard, M. J. and Cohen, P. (1989). *European Journal of Biochemistry*, **180**, 457.
16. Hubbard, M. J. and Cohen, P. (1989) *European Journal of Biochemistry*, **186**, 701.

17. Weiel, J. E., Ahn, N. G., Sager, R., and Krebs, E. G. (1990). *Advances in Second Messenger and Phosphoprotein Research*, **24**, 182.
18. Edelman, A. M., Blumenthal, D. K., and Krebs, E. G. (1987). *Annual Reviews in Biochemistry*, **56**, 567.
19. Beebe, S. Y. and Corbin, J. D. (1986). *The Enzymes*, **17**, 44.
20. Taylor, S. S., Bubis, J., Toner-Webb, J., Saraswat, L. D., First, E. A., Buechler, J. A., Knighton, D. R., and Sowadski, J. (1988). *Federation of American Societies for Experimental Biology Journal*, **2**, 2677.
21. Branson, H. N., Kaiser, E. T., and Mildvan, A. S. (1984) *CRC Critical Reviews in Biochemistry*, **15**, 93.
22. Lincoln, T. M. and Corbin, J. D. (1983). *Advances in Cyclic Nucleotide Research*, **15**, 139.
23. Sellers, J. R. and Adelstein, R. S. (1986). *The Enzymes*, **18**, 382.
24. Stull, J. T., Nunnally, M. H., and Michinoff, C. H. (1986). *The Enzymes*, **17**, 114.
25. Cohen, P. (1988). *Molecular Aspects of Cell Regulation*, **5**, 145.
26. Nairn, A. C. Hemmings, H. C., Jr., and Greengard, P. (1987). *Annual Reviews in Biochemistry*, **54**, 931.
27. Schulman, H. (1988). *Advances in Second Messenger Phosphoprotein Research*, **22**, 39.
28. Colbran, R. J., Schworer, C. M., Hashimoto, Y., Fong, Y. L., Smith, M. K., and Soderling, T. R. (1989). *Biochemical Journal*, **258**, 313.
29. Nairn, A. C. and Greengard, P. (1987). *Journal of Biological Chemistry*, **262**, 7273.
30. Nairn, A. C., Bhagat, B. and Palfrey, H. C. (1985). *Proceedings of the National Academy of Sciences of the USA*, **82**, 7939.
31. Nishizuka, Y. (1986). *Science*, **233**, 305.
32. Bell, R. M. (1986). *Cell*, **45**, 631.
33. Ashendel, C. L. (1985) *Biochimica et Biophysica Acta*, **822**, 219.
34. Huang, K.-P. (1991). *Current Topics in Cell Regulation*. (In press.)
35. Hathaway, G. M. and Traugh, J. A. (1982). *Current Topics in Cell Regulation*, **21**, 101.
36. Pinna, L. A., Meggio, F., Donella-Deana, A. and Brunati, A. M. (1985). *Proceedings of the 16th FEBS Congress P+A*, p. 155.
37. Orchinnikov, Y. A. (1982). *FEBS Letters*, **148**, 179.
38. Aton, B. R., Litman, B. J., and Jackson, M. L. (1984). *Biochemistry*, **23**, 1737.
39. Somers, R. L. and Klein, D. C. (1984). *Science*, **226**, 182.
40. Benovic, J. L., Strasser, R. H., Caron, M. G. and Lefkowitz, R. J. (1986). *Proceedings of the American Academy of Sciences of the USA*, **83**, 2797.
41. Roskoski (1983). In *Methods in Enzymology*, Vol. 99 (ed. J. D. Corbin and J. G. Hardman), p. 3. Academic Press, New York.
42. Lai, Y., Nairn, A. C. and Greengard, P. (1986). *Proceedings of the American Academy of Sciences of the USA*, **83**, 4253.
43. Dent, P., Campbell, D. G., Hubbard, M. J., and Cohen, P. (1989). *FEBS Letters*, **248**, 67.
44. Cohen, P. (1988). *Proceedings of the Royal Society, London*, **B234**, 115.
45. Newton, D. L., Krinks, M. H., Kaufman, J. B., Shipoach, J., and Klee, C. B. (1988). *Preparative Biochemistry* **18**, 247.

46. Czernik, A. J., Pang, D. T., and Greengard, P. (1987). *Proceedings of the American Academy of Sciences of the USA*, **84**, 7518.
47. Cohen, P. (1989). *Annual Reviews in Biochemistry*, **58**, 453.
48. Cohen, P. and Cohen, P. T. W. (1989). *Journal of Biological Chemistry*, **264**, 21435.
49. Klee, C. B., Draetta, G. F., and Hubbard, M. J. (1988). *Advances in Enzymology*, **61**, 149.
50. Klee, C. B. and Cohen, P. (1988). *Molecular Aspects of Cell Regulation*, **5**, 225.
51. Blumenthal, D. R., Takio, K., Hansen, R. S., and Krebs, E. G. (1986). *Journal of Biological Chemistry*, **261**, 8140.
52. Peters, K. A., Demaille, J. G., and Fischer, E. H. (1977) *Biochemistry*, **16**, 5691.
53. Manalan, A. S. and Klee, C. B. (1983). *Proceedings of the National Academy of Sciences of the USA*, **80**, 4291.
54. Politino, M. and King, M. M. (1987). *Journal of Biological Chemistry*, **262**, 10109.
55. Klee, C. B., Krinks, M. H., Manalan, A. S., Cohen, P., and Stewart, A. A. (1983). In *Methods in Enzymology*, Vol. 102, (ed. A. R. Means and B. W. O'Malley), p. 227. Academic Press, New York.
56. Krinks, M. H., Klee, C. B., Pant, H. C., and Gainer, H. (1988). *Journal of Neuroscience*, **8**, 2172.
57. Hubbard, M. J. and Klee, C. B. (1987). *Journal of Biological Chemistry*, **262**, 15062.
58. Hubbard, M. J. and Cohen, P. (1989). *European Journal of Biochemistry*, **186**, 711.
59. Nigg, E. A., Eppenberger, H. M. and Dutly, F. (1985). *EMBO Journal*, **4**, 2801.
60. Hiraga, A. and Cohen, P. (1986) *European Journal of Biochemistry*, **161**, 763.
61. Poulter, L., Ang, S.-G., Gibson, B. W., Williams, D. H., Holmes, C. F.B., Caudwell, F. B., Pitcher, J., and Cohen, P. (1988). *European Journal of Biochemistry*, **175**, 497.
62. MacKintosh, C., Campbell, D. G., Hiraga, A., and Cohen, P. (1988) *FEBS Letters*, **234**, 189.
63. Hemmings, H. C. Jr., Nairn, A. C., McGuinness, T. L., Huganir, R. L., and Greengard, P. (1989). *Federation of American Societies for Experimental Biology Journal*, **3**, 1583.
64. Kennedy, M. B. (1989). *Cell*, **59**, 777.
65. Abrams, T. W. and Kandel, E. R. (1988). *Trends in Neuroscience*, **11**, 128.
66. Wang, J. K., Walaas, S. I., and Greengard, P. (1988) *Journal of Neuroscience*, **8**, 281.
67. Chad, J. and Eckert, R. (1986). *Journal of Physiology*, **378**, 31.
68. Malinow, R., Schulman, H., and Tsien, R. W. (1989). *Science*, **245**, 862.
69. Malenka, R. C., Kauer, J. A., Perkel, D. J., Mauk, M. D., Kelly, P. T., Nicoll, R. A., and Waxham, M. N. (1989). *Nature*, **340**, 554.
70. Armstrong, D. L. (1989). *Trends in Neuroscience*, **12**, 117.
71. King, M. M., Huang, C. Y., Chock, P. B., Nairn, A. C., Hemmings, H. C., Chan, K. F. J. and Greengard, P. (1984). *Journal of Biological Chemistry*, **259**, 880.
72. Pallen, C. J. and Wang, J. H. (1985). *Archives in Biochemistry and Biophysics*, **237**, 381.
73. Tonks, N. K., Charbonneau, H., Diltz, C. D., Kumar, S., Circirelli, M. F.,

Krebs, E. G., Walsh, K. A., and Fischer, E. H. (1989). *Advances in Protein Phosphatases*, **5**, 149.
74. Tonks, N. K., Diltz, C. D., Fischer, E. H. (1988). *Journal of Biological Chemistry*, **263**, 6722.
75. Hidaka, H., Inagaki, M., Nishikawa, M., and Tanaka, T. (1988). In *Methods in Enzymology*, Vol. 159. (ed. J. D. Corbin and R. A. Johnson), p. 652. Academic Press, Orlando, Florida.
76. Manalan, A. S. and Klee, C. B. (1984). *Advances in Cyclic Nucleotide and Protein Phosphorylation*, **18**, 227.
77. Doonan, J. H. and Morris, N. R. (1989). *Cell*, **57**, 987.
78. Hinricksen, R. D., Burgess-Cassler, A., Soltvelt, B. C., Hennessey, T., and Kung, C. (1988). *Science*, **233**, 503.

MOLECULAR BIOLOGICAL APPROACHES

8

The expression of neurotransmitter receptors and ion channels in *Xenopus* oocytes

JOHN P. LEONARD and TERRY P. SNUTCH

1. Introduction

The microinjection of exogenous RNAs into *Xenopus* oocytes allows the examination of a variety of types of excitability proteins in a standard test environment. These include ion channels, neurotransmitter receptors, pumps, and transporters (1–5). In addition to expression studies, the oocyte system has proven useful for the isolation of cDNAs encoding these molecules [including a kainate receptor/channel (6), the serotonin-1C receptor (7, 8), and a delayed rectifier K channel (9)]. Furthermore, the oocyte system is a popular choice for studying the structural basis of receptor and channel function using mutagenized cDNAs (10–12). The ability of *Xenopus* oocytes to efficiently and accurately translate foreign RNA has been described in a previous volume in this series (36). This chapter will discuss the necessary requirements to successfully express ion channels and neurotransmitter receptors in *Xenopus* oocytes. These requirements include the need for:

- intact RNA encoding the receptor or channel;
- proper care and handling of oocytes; and
- electrophysiological methods tailored to reveal the receptor or channel of interest.

The oocyte performs properly most post-translational processing steps including glycosylations, acetylations, proteolytic cleavages and transport to the plasma membrane. This results in a reconstitution of the native physiological and pharmacological properties for a wide variety of channels and receptors (2, 13) (*Table 1*). However, there are a few channels and receptors that are either not expressed or show abnormal functional properties in oocytes (*Table 2*). For the eel sodium channel, the failure of functional expression is due to improper postranslational processing (14).

Table 1. Expression of exogenous channels and receptors in oocytes

Voltage-gated		Ligand-gated
K^+	A type delayed rectifier inward rectifier	Kainate (Na^+/K^+)
		NMDA ($Na^+/K^+/Ca^{2+}$) $GABA_A$ (Cl^-)
Ca^{2+}	DHP sensitive and insensitive	Glycine (Cl^-)
Na^+	Rat brain I, IIa, IIb, III heart, muscle	Acetylcholine ($Na^+/K^+/Ca^{2+}$) (muscle and neuronal nicotinic)

Neurotransmitter receptors	
Neurotensin	Substance K
Substance P	Bombesin
Acetylcholine–muscarinic (M1, M2)	Vasopressin
Serotonin–$5HT_{1C}$, $5HT_2$	Angiotensin II
Somatostatin	Quisqualate

Table 2. Oocyte expression failures

Channel/receptor type	Problem	Refs.
Eel sodium channel	Improper processing and lack of functional expression	14
Skeletal muscle DHP receptor	No functional expression	15
Shaker $I_{k(A)}$	Abnormal pharmacology	16
Serotonin 1A	No functional expression	17, 18

Overall, cases of successful expression are considerably more common than failures.

2. RNA synthesis and purification

2.1 General considerations

The isolation of intact RNA is the primary obstacle in the path to the successful expression of exogenous receptors and ion channels in oocytes. RNA can either be purified from a tissue or cell line which is known to express the desired receptor or ion channel, or alternatively, if a full-length cDNA is available, RNA can be synthesized *in vitro*. The requirement for intact RNA is easily met when RNA is produced by *in vitro* transcription from cDNA clones. However, for many channels and receptors, at present it is necessary to isolate RNA directly from the tissue or cell line of interest. Two general points should be made concerning receptor and ion channel RNAs.

First, most receptors and ion channels are encoded by relatively rare RNAs, and second, these RNAs can be as large as 10 to 12 kb in size [for example, those encoding voltage-gated sodium (19) and calcium channels (20)]. For these reasons it is necessary to minimize the chance of RNA degradation during purification. Degradation of RNA can result from both enzymatic (ribonuclease) and chemical agents. Chemical degradation of RNA can occur at any stage during purification, storage or handling and can be minimized by observing a few simple precautions:

- The presence of polyvalent cations catalyses the hydrolysis of RNA (especially at high temperatures). Avoid exposure of RNA to solutions containing magnesium, zinc, copper, and manganese. To chelate any contaminating polyvalent cations, it is prudent to include 1–2 mM EDTA in solutions coming in contact with purified RNA.

- Avoid extremes in pH since exposure to either high or low pH can cause RNA hydrolysis. The pH of RNA solutions should generally be kept between pH 5 and pH 8.

- Avoid high temperatures as much as possible. In general, keep RNA stocks frozen at −80°C. During experiments keep the RNA on ice until ready to use.

- Use high quality, ultrapure reagents. Phenol should be redistilled and then saturated with autoclaved 0.05 M Tris pH 7.5; formamide should be deionized with a mixed bed resin.

Enzymatic degradation of RNA is most likely to occur during the initial breakage of cells when nucleases are released. The inhibition of endogenous ribonucleases is most effectively accomplished by lysing the sample in the presence of strong chaotropic agents (such as guinidine hydrochloride, guanidinium thiocyanate or a mixture of urea and lithium chloride). In most protocols RNA is further purified by a combination of extensive deproteinization and the selective separation of RNA from contaminating DNA, proteins and polysaccharides.

Once the RNA has been purified, enzymatic degradation can still occur if exogenous ribonuclease is introduced. The following steps can significantly reduce exogenous ribonuclease contamination:

- For all manipulations use only sterile diposable plastic-ware and glassware that has been baked (overnight at 180°C).

- When making solutions use glass distilled water and autoclave where appropriate. It is usually not necessary to treat glassware or solutions with diethylpyrocarbonate.

- Wear gloves when handling RNA.

- Assume that your pH meter is contaminated with ribonuclease. All stock solutions that are used with purified RNA should therefore be pH'd by adding acid or base, mixing, and then testing a small aliquot with the pH meter. After determining the pH of the aliquot adjust the pH of the stock solution accordingly with acid or base. Recheck another aliquot with the pH meter. Since this method is tedious, to speed up subsequent preparations, record the volume of acid or base required for the correct pH.

- Keep a separate micropipetting device (Pipetman, Eppendorf, etc.) for manipulations involving small volumes of purified RNA. Do not use any such device which has been previously been in contact with solutions containing ribonuclease (e.g. used for DNA miniprep procedures).

2.2 RNA isolation

The successful expression of receptors and ion channels in oocytes requires the isolation of RNA which is both biologically active and of high molecular weight. For this purpose there is a wide range of available RNA isolation protocols from which to choose. Basically, the different procedures vary with respect to the degree with which they inhibit ribonucleases and in the method by which they separate RNA from other cellular constituents. *Protocol 1* uses a combination of high concentrations of LiCl and urea to denature proteins. In addition, LiCl selectively precipitates RNA over DNA and proteins (21). The procedure results in the isolation of high molecular weight RNA that translates efficiently in oocytes. We have found that this protocol works well for a variety of tissues and cell lines (including brain, heart, skeletal muscle, PC12 cells, and BC3H-1 cells). Cellular DNA must be sheared effectively during homogenization to avoid being coprecipitated. Some fibrous tissues, such as skeletal muscle, require that homogenization be performed with a polytron or blender. For small amounts of tissue or where a low yield of RNA is expected, it may be necessary to decrease some of the solution volumes listed in *Protocol 1*. As an alternative, the RNA isolation protocol of Chirgwin *et al.* (22) also yields high molecular length RNA that translates well in oocytes (23).

The above protocols result in the isolation of total cellular RNA which can be used directly for expression studies in oocytes. If the total RNA does not give an adequate electrophysiological response or if the RNA is to be used for fractionation studies or to synthesize a cDNA library, then poly(A)+ RNA should be isolated. Poly(A)+ RNA can be isolated by chromatography on oligo(dT) cellulose (type III, Collaborative Research) using standard binding and elution protocols (24). Using oligo(dT) cellulose, we find that a greater yield of high molecular weight RNA is obtained by first pre-treating the column with either a poly(A)-fraction from a previous run or by first passing 5–10% of the total RNA (1 mg ml^{-1}) over the column. The column is subsequently washed with elution buffer and then re-equilibrated with binding buffer prior to application of the remaining total RNA sample. After use, columns are stored at 4°C in the presence of 0.02% sodium azide and are never washed with NaOH (25). Alternatively, poly(A)+ RNA can be isolated using poly(U) Sepharose (Pharmacia). Poly(A)+ rat brain RNA isolated with poly(U) Sepharose contains approximately 2–3 times as much high molecular length RNA (> 5 kilobases) than does poly(A)+ RNA isolated using oligo(dT). If only a small amount of total RNA is available, it may be advisable to use a batch procedure for poly(A)+ RNA isolation (24).

Protocol 1. RNA purification

1. Homogenize 1 g of tissue extensively (35–45 strokes) in a pre-cooled 7-ml Douce homogenizer containing 7 ml of the following ice-cold solution:

    6 M urea
    3 M LiCl
    10 mM Na acetate pH 5.0
    0.5% sarkosyl

2. Bring the final volume up to 10 ml per gram of tissue and incubate on ice overnight.

3. Centrifuge at 12 000 r.p.m. at 4°C for 30 min using sterile centrifuge tubes.

4. Carefully pour off the supernatant, wipe the sides of the tube with a tissue and dissolve the pellet in 2 ml of cold 10 mM Tris–HCl pH 7.5/1 mM EDTA/0.5% sarkosyl per gram of starting tissue.

5. Extract the solution with 0.5 vol. of buffer saturated phenol:chloroform:isoamyl alcohol (25:24:1). Separate the phases by centrifugation for 5 min at 5000 r.p.m. at 4°C. Transfer the upper aqueous layer to a fresh tube and repeat the extraction twice more.

6. Extract the solution once with 0.5 vol. of chloroform:isoamyl alcohol (24:1). Centrifuge as in step 5 and transfer the aqueous layer to a clean tube.

7. Precipitate the RNA by addition of 2.5 vol. of 95% EtOH/0.1 vol. of 2.5 M Na acetate pH 5.0 and incubate at −20°C for 2 h to overnight.

8. Centrifuge at 12 000 r.p.m. for 10 min.

9. Discard the supernatant, was the pellet with 80% ethanol and spin as in step 8.

10. Dry the pellet and on ice, dissolve the RNA in 1–2 ml of sterile H_2O per gram of starting tissue. At this point the RNA is clean enough to perform Northern blot analysis and to isolate poly(A)+ RNA.

11. For injection into *Xenopus* oocytes, reprecipitate the RNA once more as in steps 7–9. Dry the pellet and take up in H_2O at a concentration of 2 mg ml^{-1}.
12. The RNA should be stored in aliquots at –80°C.

2.3 Synthetic RNA

Synthetic RNA derived from cDNA clones has been shown to direct the expression of functional receptors and ion channels in *Xenopus* oocytes.

A variety of plasmid cloning vectors are available which contain the promoter regions for the phage T_7, T_3, or SP_6 RNA polymerases. Vectors which have promoter regions on either side of the cDNA insertion site offer the flexibility of generating both sense and antisense RNAs from the same clone (for example, pGEM vectors from Promega and pBluescript vectors from Stratagene). To generate full-length coding transcripts the plasmid must be linearized downstream of the cDNA termination codon, either in the 3′ non-translated region or in the vector. If at all possible, use a restriction enzyme that leaves either a 5′ overhang or a blunt end as spurious transcripts can result from plasmids linearized with enzymes that leave a 3′ overhang (26). It has been observed that specific 5′ sequences, including upstream ATGs and hompolymer tails, inhibit the translation of synthetic RNA in oocytes (27, 28). For these reasons, it is prudent to remove most of the 5′ non-coding sequences of the receptor or channel cDNA during plasmid construction.

Under optimal conditions the *in vitro* transcription reaction is highly efficient, resulting in the synthesis of 8–10 mol of RNA per mol of DNA template in 1 h ((29), *Protocol* 2). Unless the 5′-end of the synthetic RNA is protected by a GpppG cap, it will be rapidly degraded upon injection into oocytes (30). Capping can be performed enzymatically after RNA transcription using guanyltransferase (30). Alternatively, a simpler protocol is to include a tenfold molar excess of (5′)GpppG(5′) to GTP in the transcription reaction (31, 32). Greater than 95% of the newly synthesized transcripts have a 5′ terminal cap under these conditions. It is not necessary for the synthetic RNA to have a poly(A) tail, although the long term stability of RNA in oocytes is increased (30, 31).

Protocol 2. Synthesis of synthetic RNA

1. In a sterile microfuge tube mix the following at room temperature:

    5 µl   10 × transcription buffer
    5 µl   5 mM ATP
    5 µl   5 mM UTP
    5 µl   5 mM CTP

    5 µl   5 mM GpppG cap
    1 µl   5 mM GTP
    1 µl   0.5 M DTT
    1.6 µl RNasin (30 units ml^{-1})
    17.4 µl H$_2$O
    2 µl   linearized plasmid (2 µg/µl)
    2 µl   30 units RNA polymerase (T$_7$, T$_3$, SP$_6$)

2. Incubate at 37°C for 60 min.
3. Add 1 µl of DNase I (RNase free; 1000 units ml^{-1} or 1 mg ml^{-1}).
4. Incubate at 37°C for 10 min.
5. Extract with 25 µl of phenol:chloroform:isoamyl alcohol (25:24:1).
6. Vortex lightly and spin in microfuge for 3 min.
7. Transfer the supernatant to a fresh tube and bring the volume up to 100 µl with sterile H$_2$O.
8. Separate the RNA transcripts from unincorporated nucleotides by spin column chromatography.
9. Precipitate RNA transcripts by addition of 2.5 vol. 95% EtOH and 0.1 vol. 2.5 M Na acetate pH 5.0 and incubate at −20°C for 2 h to overnight.
10. Pellet RNA in microfuge for 10 min.
11. Wash with 400 µl of 80% EtOH and spin as in step 10. Dry pellet and take up in sterile H$_2$O at a concentration of 0.1 to 0.25 µg ml^{-1}.

Comments:

10 × transcription buffer:
  200 mM Tris–HCl pH 7.5
  100 mM NaCl
  60 mM MgCl$_2$
  20 mM spermidine

Nucleotides: stocks are made up in 70 mM Tris–HCl pH 7.5

To monitor synthesis, include 1 µl of α-^{32}P-CTP (approx. 400 Ci mol^{-1}).

2.4 RNA size fractionation

The characterization of a particular receptor or channel in oocytes injected with tissue or cell line RNA can be complicated by the fact that the investigator cannot control the co-expression of other conductances encoded by the RNA. To help circumvent this problem, and also to increase the signal size, the RNA can be fractionated according to size prior to injection (33).

Using this approach, individual fractions or pools of fractions are injected into oocytes and those which encode the receptor or channel of interest are identified. In addition to enriching for the RNA of interest, size fractionation allows reconstitution experiments to be performed. For example, rat brain RNAs which are smaller than those which encode the alpha subunit of sodium channels have been shown to modulate the inactivation properties of sodium channels expressed in oocytes (34). Size fractionation of RNA and the identification of an active fraction is also a prerequisite to the construction of enriched cDNA libraries.

Most conveniently the size fractionation of RNA is accomplished by centrifugation through a preformed sucrose gradient (*Protocol 3*). A 5–20% sucrose gradient is effective in separating RNAs in the 2 to 12 kb size range. In order to minimize RNA aggregation, gradients are performed under low salt conditions. In addition, to minimize RNA degradation the gradients are carried out at 4°C (33, 34). To recover a given receptor or channel activity in as small a size range as possible, the poly(A)+ sample should be loaded onto the gradient in a small volume (less than 100 µl). Fractions are recovered by inserting an 18-gauge needle into the bottom of the centrifuge tube and dripping 400 µl samples into a 1.5-ml microfuge tubes (numbered and set up in advance). After recovery of the RNA by ethanol precipitation the fractions can be used directly for oocyte injection, or alternatively they can be pooled and reprecipitated. The amount of RNA in each fraction can be determined by a spot test using ethidium bromide as described for DNA by Maniatis *et al.* (24). Alternatively, each fraction can be taken up in a defined volume of water and a specific activity defined (e.g. poly(A)+ µg equivalents per microlitre; *Protocol 3*).

In order to determine the RNA size distribution in each fraction and to estimate the degree of enrichment, a portion of each sample is separated through a denaturing agarose gel and assayed by Northern blot (*Protocols 4 and 5*). Hybridization of the gel blot to a ^{32}P-TTP probe permits both quantitative and qualitative determinations of the fractionated poly(A)+ RNA (*Figure 1*; also ref. 25)). Typically, using rat brain RNA a 50- to 60-fold molar enrichment is obtained for RNAs greater than 7 kb. The degree of enrichment for RNAs in the 2–4 kilobase range is significantly lower.

Protocol 3. Sucrose density gradient preparation.

1. Reprecipitate 150 µg of poly(A)+ RNA and take up in sterile 1 mM EDTA, pH 7.5 at a concentration of 2 µg µl^{-1}
2. Using a standard gradient maker, prepare linear 5–20% sucrose gradients in polyallomer tubes. Gradient solutions are composed of the following:
    (a) 5% sucrose

15 mM Pipes pH 6.5
5 mM EDTA
0.25% sarkosyl

(b) 20% sucrose
15 mM Pipes pH 6.5
5 mM EDTA
0.25% sarkosyl

The solution should be filter sterilized. The best resolution will result from using long-bucket swing out rotors (for example, the Beckman SW28.1 with 14 × 89 mm tubes). When using a peristaltic pump to prepare the gradients, the tubing should first be acid washed and then rinsed exhaustively with H_2O.

3. Precool the gradients and rotor at 4°C for 1 h.
4. Heat the RNA sample at 75°C for 3 min, cool on ice.
5. Carefully layer the sample on to the gradient.
6. Centrifuge the gradients at 80 000 g for 18 h at 4°C with the Beckman 28.1 rotor or equivalent. Under these conditions 9–10 kilobase RNAs will sediment approximately 70% down the tube (*Figure 1*).
7. Collect 400 μl fractions into 1.5-ml microfuge tubes by inserting a needle into the bottom of the centrifuge tube. Add 40 μl of 2.5 M sodium acetate, pH 5.0 and fill the remainder of the microfuge tube with cold ethanol. Place at −20°C overnight.
8. Recover the RNA by centrifugation, wash twice with 400 μl of 80% ethanol, dry and take up each fraction in an identical volume of sterile H_2O (25–75 μl). Samples should be stored at −80°C. The recovery of RNA from the gradients is nearly 100%. Therefore, if 150 μg of poly(A)+ RNA was loaded on to the gradient and each fraction taken up in 75 μl of H_2O, the concentration in each fraction is 2 μg equivalents of poly(A)+ RNA $μl^{-1}$. The fractions can be directly injected into oocytes or alternatively, various fractions can be pooled and re-pricipitated prior to injection.

Protocol. 4. Estimation of mRNA sizes of sucrose gradient fractions: preparation of ^{32}P-poly(dT) probe

1. Mix the following at room temperature:

    5 μl 10 × cacodylate buffer (1 M sodium cacodylate/250 mM Tris pH 7.6)
    5 μl 1 mM DTT

Expression of receptors and channels in Xenopus oocytes

    1.5 μl poly(dT$_{15}$) (10 μg ml^{-1})
    10 μl α-^{32}P-TTP (3000 Ci mol^{-1})
    26 μl H$_2$O

2. Incubate at 37°C for 10 min.
3. Add 1.5 μl of 50 mM CoCl$_2$ and 60 units of terminal transferase.
4. Incubate for 60 min at 37°C.
5. Add 5 μl of cold TTP (10 mM) and 60 units of terminal transferase.
6. Incubate at 37°C for 8–10 min.
7. Add 2 μl of 250 mM EDTA and purify the poly(dT) probe by spin column chromatography using Sephadex G-50 (Pharmacia) equilibrated with 10 mM Tris pH 7.5/1 mM EDTA.
8. The probe can be used directly from the spin column elutant or can be stored at −20°C for 1–2 weeks prior to use.

Protocol 5. Agarose gel fractionation and Northern blot of gradient fractions

1. Pour a 1.1% agarose gel containing 2.2 M formaldehyde and 1 × Mops buffer (in a fume hood!).
2. On ice mix the following:
        5 μl 10 × Mops buffer
        8.75 μl formaldehyde
        25 μl formamide (deionized)
            variable sucrose gradient fraction (use approximately 5 μg equivalent of poly(A)+ RNA – e.g. 1/30th of a 150 μg gradient)
        variable H$_2$O to 50 μl final volume
3. Heat sample to 65°C for 15 min and then cool on ice.
4. Add 2 μl of loading buffer (50% glycerol/1 mM EDTA/0.4% Bromophenol Blue/0.4% xylene cyanol/10% Ficoll).
5. Before loading samples rinse wells with 1 × Mops running buffer.
6. Load sample and run at approximately 80 V until the Bromophenol Blue reaches the bottom of the gel.
7. Carefully remove gel and take a picture of the markers.
8. Rinse out the gel box with H$_2$O and fill the chambers approximately half-full with 20 × SSPE. Transfer the RNA by capillary blot overnight to a nylon membrane such as Hybond-N (Amersham). Fix the RNA to the filter by UV cross-linking as described by the manufacturer.

9. Pre-hybridize the filter at 42°C for 60 min in the following solution:
   90 mM NaCl
   10 mM sodium acetate pH 6.0
   10 mM EDTA
   0.1% sodium pyrophosphate
   0.2% SDS
   500 µg ml^{-1} heparin
10. Discard the pre-hybridization buffer and add a fresh aliquot. Heat the radiolabelled poly(dT) probe to 65°C for 5 min and hybridize at 42°C for 14–18 h. Use between 2–4 × 10^5 c.p.m. probe/ml hybridization buffer.
11. Wash the filter four times 20 min in 0.025 M NaCl/0.1% SDS. Air-dry and expose to X-ray film. If 3 to 5 µg equivalents of poly(A)+ RNA were loaded per lane, the filter will only require a few hours' exposure.
12. Using the exposed autoradiogram, the relative amount of RNA in each sucrose gradient fraction can be assayed. The individual lanes of the autoradiogram can be scanned with a densitometer and the area under each peak determined. Alternatively, using the autoradiogram as a guide, the individual lanes can be cut from the hybridization filter and the amount of ^{32}P bound determined by scintillation counting.

3. Preparation and injection of oocytes

3.1 *Xenopus* frogs and oocytes

Xenopus laevis are available from several companies including: *Xenopus* I, Inc., (Ann Arbor, Michigan, USA), Nasco Biologicals, Inc. (Fort Atkinson, Wisconsin, USA), and *Xenopus* Ltd, (Nutfield, UK). Avoid shipments during hot weather as the frogs are traumatized, resulting in oocytes of poor quality. Mature females that have been injected with human chorionic gonadotropin (HCG) are recommended because the ovaries are proven functional and a fresh cycle of oocyte development is induced. Female frogs usually produce healthy oocytes if they are kept at 17–19°C, on a regular dark/light cycle (e.g. 12 h dark/12 h light) and fed beef heart or liver twice a week (about 8 g/frog) (35). *Xenopus* will not eat frog brittle unless raised on it. The tanks are washed with water twice per week after the frogs have eaten. Fresh water must be free of chlorine and fluorine before returning the frogs to the cleaned tanks. Dehalogenation is accomplished by bubbling air through the water for several hours or allowing the water to stand overnight. When healthy frogs cease to produce healthy oocytes, a new shipment of frogs may solve the problem.

Oocytes are isolated from frogs anaesthetized in 0.2% 3-aminobenzoic acid ethyl ester (Tricaine methane sulphonate or MS-222). Treat all surgical

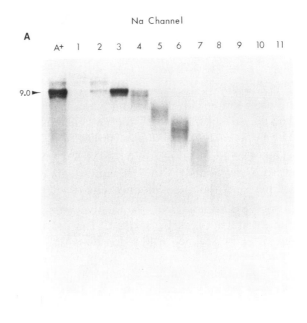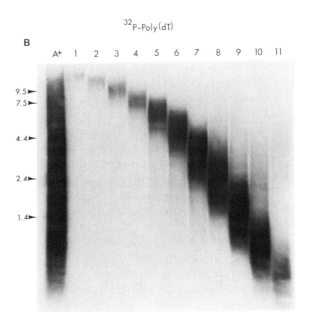

Table 3. Composition of solutions

Component:	OR-2 (mM)	Incubation (mM)	SOS (mM)	OSOP (mM)	I_{Na} (mM)	I_K (mM)	I_{Ba} (mM)	I_{NMDA} (mM)
NaCl	82.5	100.0	100.0	10.0	100.0	100.0	–	100.0
KCl	2.0	2.0	2.0	90.0	2.0	2.0	2.0	2.0
$MgCl_2$	1.0	1.0	1.0	–	1.0	2.8	–	–
$CaCl_2$	–	1.8	1.8	–	1.8	–	–	1.8
Hepes	5.0	5.0	5.0	10.0	5.0	5.0	5.0	5.0
$BaCl_2$	–	–	–	–	–	–	40.0	–
Gentamycin	–	10 µg ml^{-1}	–	–	–	–	–	–
Na-pyruvate	–	2.5	–	–	–	–	–	–
EGTA	–	–	–	10.0	–	–	–	–
Tetraethyl-ammonium	–	–	–	–	10.0	–	36.0	–
4-Aminopyridine	–	–	–	–	5.0	–	5.0	–
Niflumic acid	–	–	–	–	–	–	0.2	0.2
pH	7.5	7.5	7.5	7.4	7.5	7.5	7.6	7.5

Comments:
1. Or-2 is Ca-free saline for defolliculation.
2. Incubation medium is used to culture oocytes.
3. SOS is 'standard oocyte solution' for routine recording purposes.
4. OSOP is 'outside-out patch' solution.
5. 'I_{Na}' is for recording I_{Na}.
6. 'I_K' is for recording I_K.
7. 'I_{Ba}' is for recording I_{Ba} through Ca channels.
8. 'I_{NMDA}' is for recording I_{NMDA} currents.
9. Niflumic acid (from Sigma) is a potent reversible inhibitor of the $I_{Cl(Ca)}$ in oocytes (40).

instruments with 70% ethanol prior to use. A 1–2 cm incision is made in the abdomen and several ovarian lobes are removed (36). To prevent acidification of the frog's skin by the anaesthetic, rinse the skin frequently with water during the surgery. Use surgical silk to suture the incision closed. For recovery from anaethesia, place the frog in a small tank with shallow water. Keep recovered, post-operative frogs in a special tank. The wounds heal very

Figure 1. Gel blots of sucrose-gradient fractionated rat brain RNA. Poly(A)+ RNA (150 µg) was separated according to size through a linear 5–20% sucrose gradient and 33 fractions were collected as described in *Protocol 3*. Five microgram equivalents of poly(A)+ RNA from sets of three adjacent fractions were pooled (**Lanes 1** through **11** in panels **A** and **B**), separated through an agarose gel containing 2.2 M formaldehyde and blotted to Hybond-N as described in *Protocol 5*. The blot in panel **A** was hybridized to a ^{32}P-labelled Na channel probe, while the blot in panel **B** was hybridized to a ^{32}P-poly(dT) probe (*Protocol 4* and *5*). A+ in each panel denotes 5 µg of unfractionated poly(A)+ RNA. Upon injection into *Xenopus* oocytes, only those pools which contained the intact 9.0 Kb sodium channel RNA resulted in the appearance of functional Na channels. The smearing in panel A (pools 4–8) is the result of sodium channel RNA degradation.

quickly and cleanly due to novel antibiotics present in the frogs' skin (37). Each ovary contains thousands of oocytes, hence oocytes can be isolated as needed from each frog a number of times.

Using forceps the isolated ovarian lobes are gently stripped of connective tissue releasing small clumps of oocytes. The oocytes are denuded of overlying follicle cells by gentle agitation in 2 mg ml^{-1} collagenase (Sigma type I-A) in a Ca-free soution (OR-2) (see *Table 3*). The collagenase treatment should continue until 50% of the oocytes ar defolliculated (about 1.5–3 h) to avoid overdigestion. Many of the non-denuded oocytes are defolliculated during the wash as the follicle cell sac sticks to the bottom of the culture dish. After collagenase treatment wash the oocytes thoroughly in standard oocyte saline (SOS), see *Table 3*. If contamination of the oocytes does occur, add 100 units ml^{-1} penicillin and 100 µg ml^{-1} streptomycin. Do not use the oocytes for injection of RNA for at least 2 h post-defolliculation (38).

Selection of oocytes should be done immediately after defolliculation. This is an often overlooked but crucial step. The ideal cell is a stage V or VI oocyte (39) and of uniform pigmentation over the animal pole when viewed under a dissecting microscope at 50×. If during the initial incubation many of the freshly isolated oocytes may die prematurely. Oocytes to be avoided include incompletely denuded cells and those with dark spots, splotches, or dark/light marbling. Oocytes with a diffuse light patch covering about one-third of the animal pole should also be avoided as they are undergoing spontaneous maturation.

3.2 Injection and culture of ooctyes

An oocyte injection chamber can be constructed in a 35-mm Falcon tissue culture dish. A 0.5-mm weave, square, nylon, monofilament mesh (Small Parts, Inc., Miami, Florida, USA) cemented on to the dish with Sylgard or epoxy glue will prevent the ooctyes from rolling around. Place a large drop of water in the centre of the mesh to keep the Sylgard out of this region during curing. Injection is accomplished with the chamber positioned under a dissecting microscope. A Drummond 100-µl microdispenser mounted on a micromanipulator (Brinkmann or Narashige) with a Plexiglass collar is used to inject mRNA into the oocytes. It is essential to order extra metal rods for the microdispenser as only straight rods are effective. If a rod is bent it will auger out a hole in the oocyte. (Drummond is developing a microdispenser without a rotating rod.)

Glass microdispensor needles with very gradual shanks are pulled with a microelectrode puller and then broken into a 20-µm tip opening with a fire-polished pasteur pipette or sterile forceps. Asymmetrical or 'tounged' tips are best.

Healthy oocytes are injected with 50–70 nl aqueous samples of 1–2 µg µl^{-1}

poly(A)+ RNA or 2–4 µg µl^{-1} of total RNA. Due to difficulties in predicting the efficiency with which *in vitro* transcribed RNA will be translated *in ovo*, it is advisable to initially test several concentrations (in the range of 0.5–03 µg µl^{-1}). Always include as a positive control a group of oocytes injected with a sample of RNA with previously characterized effectiveness. It is therefore possible to assign any failure of the oocytes to properly translate and express a new RNA sample to the sample, and not to the particular batch of oocytes used in the experiment. Include as a negative control a group of oocytes injected with water.

After injection the oocytes are incubated at 20°C in sterile gentamycin and sodium pyruvate supplemented SOS (*Table 3*). The dishes are kept in a humid atmosphere and the solution is replaced at least once each day. Under these conditions the oocytes can be maintained in good health for up to 10 days. Expressed channels are observable 1–3 days after injection, and may be studied for an additional 7 days.

4. Electrophysiology of channels expressed in oocytes

4.1 Two-electrode voltage clamping

The oocyte has the ideal shape for two-electrode voltage clamping. However, the size of the oocyte results in an increase in capacitance to charge (0.5 µF), which limits clamp speed. Two available systems do allow reasonable two-electrode voltage clamping: the +/− 80-V Dagan 8500 and 8800, and the +/− 30-V Axoclamp 2-A. Dagan and Warner Instruments have recently designed new systems for oocytes (+/− 130 V) that should be faster.

Operator controlled factors that limit the speed of two-electrode voltage clamps include electrode resistance and capacitive coupling between the current and voltage electrodes. Generally, electrodes are filled with 3 M KCl and have resistances of 0.5 to 2 MΩ. Grounds are either large KCl/agar-filled glass capillaries (1 cm × 2 mm) or Ag–AgCl pellets. The KCl-Agar bridges are essential when Cl-free salines are used because of the creation of junction potentials on the Ag–AgCl pellets. Capacitive coupling is reduced by placing an insulated metal-foil ground shield in the bath between the two electrodes, and keeping the bath level as low as possible. When all these precautions are followed, the settling time of the capacitive current is 1–3 msec at room temperature (*Figure 2A*). This clamp speed is adequate for most of the currents examined. However, if further reduction is desired, the electrode resistance can be lowered to 0.2 MΩ by pulling borosilicate glass in two stages as for patch pipettes and then filling them with an intracellular-like saline (41). In addition, clamp speed can be improved significantly by using small stage II oocytes with a surface area about 10% that of stage V and VI oocytes (42). However, RNA injection of these oocytes requires a special apparatus (Nanopump by WP Instruments, or other such ultrastable device).

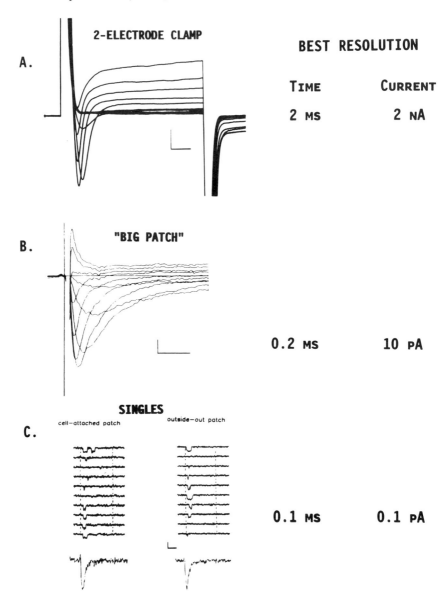

Figure 2. Examples of voltage-clamped current recordings in SOS from oocytes injected with rat brain mRNA. **A**: Standard 2-electrode voltage-clamping yields inward sodium and outward potassium and slow outward Ca-dependent Cl tail currents with time and amplitude resolution shown in the two columns on the right: **B**: Patch recording of devitellinized oocytes with a large diameter (20 μm) pipette allows tenfold better time-resolution. (K currents and Cl currents blocked with caesium and cadmium.) **C**: Patch recording with smaller diameter pipettes allows resolution of single Na channel currents. Calibration bars: **A**: 400 nA; 5 msec. **B**: 20 pA; 2 msec. **C**: 1 pA; 3 msec.

Two problems frequently encountered in routine two-electrode clamping are fast activation time-courses and tail currents that exceed the clamp speed. These cases are best studied using the 'big patch' recording technique described in Section 4.3. A further problem is the introduction of changes in the series resistance (43) in cases of highly successful ion channel expression where currents exceed 1 μA. To test for this potential artefact, partially block the current of interest or inject less RNA and examine the kinetics and I-V relationship for dependence on the current amplitude. If they are amplitude dependent, the artefact can be eliminated by injecting less RNA in subsequent groups of oocytes.

4.2 Patch clamping

The patch-clamp technique can be used on oocytes both to record macroscopic currents with improved time resolution (44, 45) and to make traditional single-channel recordings (34, 41, 45–47). In order to obtain a gigaohm seal between the oocyte surface and the electrode, the outer vitelline membrane protecting the plasma membrane must be removed. This is done by treating the oocytes in a hypertonic solution of normal saline (SOS) containing an additional 60–100 mM NaCl to partially dehydrate the cell. A gap forms between the vitelline membrane and the plasma membrane as a result (some oocytes will shrink quicker than others). When viewed under a dissecting microscope the vitelline membrane can then be removed with forceps. Avoid exposing devitellinized oocytes to air–water interfaces as they will lyse.

Patch pipettes are pulled in two stages to the desired diameter. The pipettes are then coated with Sylgard to within 100 μm of the tip and fire-polished. This treatment reduces the capacitance of the pipettes resulting in cleaner recordings.

Gigaohm seal formation for both big and small patches is achieved in an analogous manner to other cell types (48). Generally, it is more difficult to obtain gigaohm seals with large diameter pipettes than with smaller ones. Seal formation can occur quite abruptly but in the big patches usually develops over a few tens of seconds. After the pipette has touched the membrane gentle suction is applied with a micrometer syringe (2 ml Gilmont) to form the gigaohm seal. Alternatively, mouth suction is used. A good patch will remain sealed for 10–45 min.

4.3 'Big patches'

The 'big patch' clamping technique is used when time-resolution of macroscopic currents greater than that achieved with two-electrode voltage clamping is desired (*Figure 2B*). Pipette tips are made from soft glass capillary tubing (e.g. Kimble No. 73811) pulled in two stages. Typical big patch-pipette tip diameters are 20–30 μm and have resistances of about 200 kΩ in SOS.

Theoretically, tips of this diameter enclose a hemispheric surface area of roughly 500 µm^2; however, due to extensive folding of the oocyte-cell surface, the actual area enclosed is greater than 200 µm^2 (about 1×10^{-4} of the total oocyte surface area). A patch from a cell with a whole-cell current of 300 nA will average 30 pA current. Such a current was resolvable during a study of Na currents (44); however, 100 pA current is preferable. Sufficient channel density can be achieved by injecting 0.3 mg ml^{-1} to 0.6 mg ml^{-1} synthetic RNA (11) or 1.0 to 5.0 mg ml^{-1} whole brain mRNA. This has been effective for I_{Na} studies (34, 49) as the current has been shown to increase linearly with RNA in this concentration range. It is possible to titrate the density of channels on the membrane surface to any desired level if cDNA clones of the channel of interest are available. This enables the use of the easier sealing intermediate diameter patch-pipettes (45) to record macroscopic currents with excellent time resolution.

4.4 Single-channel recording

The oocyte expression system is a very good preparation to study single-channel properties (*Figure 2C*). A primary advantage is the control of expression level of the channel of interest by varying the amount of RNA injected. The expression level can be varied such that a typical pipette on average encloses a single channel. Secondly, interference from endogenous channels is minimal. Cell-attached, and to a lesser degree, inside-out, patches are occasionally subject to interference from 'stretch-activated channels'. Consequently, the outside-out patch configuration is preferred because such interference is rare (41, 50, 51).

Small patch-pipettes are made from hard glass capillary tubing (e.g. KG-33 from Garner Glass Co.) and have reisstances of 10–20 MΩ. Typical seal resistances of small patches are greater than 20 GΩ. Uncompensated capacitance and leak must be subtracted from each single-channel recording. This is generally done by averaging traces without channel openings and subtracting this average from each trace with channel openings. Alternatively, a scaling procedure can be used.

4.5 Isolation of exogenous currents

4.5.1 Endogenous activity

One constraint of any *in vivo* expression system is the need to control for any endogenous ion channel activity that may be present. Several different ion channel types are endogenous to *Xenopus* oocytes even after removal of follicle cells (see *Table 4*) (46, 52–55). Levels of endogenous channels can vary from frog to frog, but usually not between oocytes of the same stage from the same frog. For example, while calcium channels measured in barium saline may be undetectable from many frogs, an exceptional frog may produce oocytes that show a 50 nA peak barium current. This phenomenon is not

Table 4. Endogenous currents

Receptor/Agonist	Ionic conductance	Comments
Progesterone	?	no immediate effect but induces membrane depolarization over a period of several hours initiates oocyte maturation
Intracellular Ca^{2+}?	↑ gCl	depolarizing at oocyte resting potential single channel approx. 3 pS conductance with mean open time approx. 200 msec possibly one type of Ca^{2+}-dependent Cl^- channel which can be activated by receptor stimulation and release of Ca^{2+} from endogenous stores; Ca^{2+} injection; entry of Ca^{2+} through voltage-generated Ca^{2+} channels; spontaneous release of Ca^{2+} from intracellular stores whole-cell current is approx. 0–1000 nA when activated by I_{Ca}
Voltage-activated	gNa^+	Vm must be held at +20 mV or greater for > 10 seconds to remove inactivation blocked by 1 μM TTX
Voltage-activated	gCa^{2+}	whole-cell current carried by Ba^{2+} is approx. 0–20 nA transient (tau = 100 msec) peak current at +15 ± mV no effect of nifedipine
Voltage-activated	gK^+	slow non-inactivating delayed rectifier type K^+ current whole current is approx. 0–100 nA

restricted to calcium channels. Oocytes from approximately 2% of frogs show bona fide tetradotoxin-sensitive sodium currents. One solution is to increase the signal-to-noise ratio by overwhelming the small endogenous channel activity with the exogenous channel of interest. A second solution is to take advantage of clear differences in the properties of the endogenous channel and the exogenous channel. The current of interest can be dissected out using either pharmacological (*Table 3*) or voltage-clamp pulse protocols. In order to control for any effects of endogenous channels, test several non-injected oocytes in parallel with studies on injected oocytes.

4.5.2 Exogenous interfering currents

A clear advantage to using receptor and ion channel cDNAs is that RNA transcribed *in vitro* is free of interfering exogenous currents when injected with oocytes. However, when tissue derived total RNA or poly(A)+ RNA is used, the same sorts of pharmacological and physiological techniques are used that isolated the current of interest from endogenous oocyte currents. Examples of salines and drugs used to isolate particular classes of currents induced by injection of rat brain RNA are listed in *Table 3*. The voltage pulse protocols are numerous and should be designed to mimic the expected response in the tissue from which the RNA was isolated.

References

1. Jan, L. Y. and Jan, Y. N. (1989). *Cell*, **56**, 13.
2. Lester, H. A. (1988). *Science*, **241**, 1057.
3. Catterall, W. A. (1988). *Science*, **242**, 50.
4. Barnard, E. A., Miledi, R., and Sumikawa, K. (1982). *Proceedings of the Royal Society*, London, B **215**, 241.
5. Gundersen, C. B., Miledi, R., and Parker, I. (1983). *Proceedings of the Royal Society*, B **219**, 103.
6. Hollmann, M., O'Shea-Greenfield, A., Rogers, S. W., and Heinemann, S. (1989). *Nature*, **342**, 643.
7. Julius, D., MacDermott, A. B., Axel, R., and Jessel, T. M. (1988). *Science*, **241**, 558.
8. Lubbert, H., Hoffman, B. J., Snutch, T. P., van Dyke, T., Levine, A. J., Hartig, P. R., Lester, H. A., and Davidson, N. (1987). *Proceedings of the National Academy of Sciences of the USA*, **84**, 4332.
9. Frech, G. C., VanDongen, A. M. J., Schuster, G., Brown, A. M., and Joho, R. H. (1989). *Nature*, **340**, 642.
10. Sakmann, B., Methfessel, C., Mishina, M., Takahashi, T., Takai, T., Kurasaki, M., Fukuda, K., and Numa, S. (1985). *Nature*, **318**, 538.
11. StGhmer, W., Conti, F., Suzuki, H., Wang, X., Noda, M., Yahagi, N., Kubo, H., and Numa, S. (1989). *Nature*, **339**, 597.
12. Leonard, R. J., Labarca, C. G., Charnet, P., Davidson, N., and Lester, H. A. (1988). *Science*, **242**, 1578.
13. Snutch, T. P. (1988). *Trends in Neuroscience*, **11**, 250.
14. Thornhill, W. B., and Levinson, S. R. (1987). *Biochemistry*, **26**, 4381.
15. Perez-Reyes, E., Kim, H. S., Lacerda, A. E., Horne, W., Wei, X., Rampe, D., Campbell, K. P., Brown, A. M., and Birnbaumer, L. (1989). *Nature*, **340**, 233.
16. MacKinnon, R., Reinhart, P. H., and White, M. M. (1988). *Neuron*, **1**, 997.
17. Kobilka, B. K., Frielle, T., Collins, S., Yang-Feng, T., Kobilka, T. S., Franke, U., Lefkowitz, R. J., and Caron, M. C. (1987). *Nature*, **329**, 75.
18. Fargin, A., Raymond, J. R., Lohse, M. J., Kobilka, B. K., Caron, M. C., and Lefkowitz, R. J. (1988). *Nature*, **335**, 358.
19. Noda, M., Ikeda, T., Suzuki, H., Takeshima, H., Takahashi, T., Kuno, M., and Numa, S. (1986). *Nature*, **322**, 826.
20. Snutch, T. P., Leonard, J. P., Gilbert, M. M., Lester, H. A., and Davidson, N. (1990). *Proceedings of the National Academy of Sciences of the USA*, **83**, 3391.
21. Auffray, C., and Rougeon, F. (1980). *European Journal of Biochemistry*, **107**, 303.
22. Chirgwin, J. M., Przybyla, A. E., MacDonald, R. J., and Rutter, W. J. (1979). *Biochemistry*, **18**, 5294.

23. Trimmer, J. S., Cooperman, S. S., Tomika, S. A., Zhou, J., Crean, S. M., Boyle, M. B., Kallen, R. G., Sheng, Z., Barchi, R. L., Sigworth, F. J., Goodman, R. H., Agnew, W. S., and Mandel, G. (1989). *Neuron*, **3**, 33–49.
24. Maniatis, T., Fritsch, E. F., and Sambrook, J. (1982). *Molecular Cloning*. Cold Spring Harbor Laboratory, Cold Spring Harbor, NY.
25. Lubbert, H., Snutch, T. P., Dascal, N., Lester, H. A., and Davidson, N. (1987). *Journal of Neuroscience*, **7**, 1150.
26. Schenborn, E. T. and Mierendorf, R. C., Jr. (1985). *Nucleic Acids Research*, **13**, 6223.
27. Galili, G., Kawata, E. E., Cuellar, R. E., Smith, L. D., and Larkins, B. A. (1986). *Nucleic Acids Research*, **14**, 1511.
28. Kobilka, B. K., MacGregor, C., Daniel, K., Kobilka, T. S., Caron, M. C., and Lefkowitz, R. J. (1987). *Journal of Biological Chemistry*, **262**, 15796.
29. Melton, D. A., Kreig, P. A., Rebagliati, M. R., Maniatis, T., Zinn, K., and Green, M. R. (1984). *Nucleic Acids Research*, **12**, 7035.
30. Kreig, P. A., and Melton, D. A. (1984). *Nucleic Acids Research*, 12, 7057.
31. Drummond, D. R., Armstrong, J., and Colman, A. (1985) *Nucleic Acids Research*, **13**, 7375.
32. White, M. M., Mixter-Mayne, K., Lester, H. A., and Davidson, N. (1985). *Proceedings of the Natural Academy of Sciences of the USA*, **82**, 4852.
33. Sumikawa, K., Parker, I., and Miledi, R. (1984). *Proceedings of the National Academy of Sciences of the USA*, **81**, 7994.
34. Krafte, D. S., Snutch, T. P., Leonard, J. P., Davidson, N., and Lester, H. A. (1988). *Journal of Neuroscience*, **8**, 2859.
35. Brown, A. L. (1970). *The African Clawed Toad* Xenopus laevis*: A Guide for Laboratory Practical Work*. Butterworths, London.
36. Colman, A. (1984). In *Transcription and Translation: A Practical Approach* (ed. B. D. Hames and S. J. Higgins) pp. 271–302. IRL Press, Oxford.
37. Fricker, L. D. (1988). *Einstein. Quarterly Journal of Biological Medicine*, **6**, 36.
38. Dascal, N. (1987). *CRC Critical Reviews in Biochemistry*, **22**, 317.
39. Dumont, J. N. (1972). *Journal of Morphology*, **136**, 153.
40. White, M. M. and Aylwin, M. (1990). *Molecular Pharmacology*, **37**, 720.
41. Methfessel, C., Witzeman, V., Takahashi, T., Mishina, M., Numa, S., and Sakmann, B. (1986). *Pflügers Archiv für die Gesamte Physiologie*, **407**, 577.
42. Krafte, D. S. and Lester, H. A. (1989). *Journal of Neuroscience Methods* **26**, 211.
43. Finkel, A. S., and Gage, P. W. (1985). In *Voltage and patch clamping with microelectrodes* (ed. T. G. Smith, H. Lecar, S. J. Redman, and P. W. Gage), pp. 47–94. Waverly, Baltimore, Maryland.
44. Leonard, J. P., Snutch, T. P., Lubbert, H., Davidson, N., and Lester, H. A. (1986). *Biophysical Journal*, **49**, 386a.
45. Stuhmer, W., Methfessel, C., Sakmann, B., Noda, M., and Numa, S. (1987) *European Biophysical Journal*, **14**, 131.
46. Moorman, J. R., Zhou, Z., Kirsch, G. E., Lacerda, A. E., Caffrey, J. M., Lam, D. M., Joho, R. H., and Brown, A. M. (1987). *American Journal of Physiology*, **253**, H985.
47. Auld, V. J., Goldin, A. L., Krafte, D. S., Marshall, J., Dunn, J. M., Catterall, W. A., Lester, H. A., Davidson, N., and Dunn, R. J. (1988). *Neuron*, **1**, 449.

48. Hamill, O. P., Marty, A., Neher, E., Sakmann, B., and Sigworth, F. J. (1981). *Pflügers Archiv für die Gesamte Physiologie*, **391**, 85–100.
49. Sumikawa, K., Parker, I., and Miledi, R. (1986). *Progress in Zoology*, **33**, 127.
50. Guhary, F. and Sachs, F. (1984). *Journal of Physiology*, **352**, 685.
51. Yang, X.-C. and Sachs, F. (1989). *Science*, **243**, 1068.
52. Dascal, N., Snutch, T. P., Lubbert, H., Davidson, N., and Lester, H. A. (1986). *Science*, **231**, 1147.
53. Leonard, J. P., Nargeot, J., Snutch, T. P., Davidson, N., and Lester, H. A. (1987). *Journal of Neuroscience*, **7**, 875.
54. Snutch, T. P., Leonard, J. P., Nargeot, J., Lubbert, H., Davidson, N., and Lester, H. A. (1987). *Society of General Physiologists Series*. **42**, 154.
55. Umbach, J. and Gundersen, C. (1987). *Proceedings of the National Academy of Sciences of the USA*, **84**, 5464.

9

Molecular approaches to the structure and function of the GABA$_A$ receptors

F. ANNE STEPHENSON and MICHAEL J. DUGGAN

1. Introduction

The GABAergic system is ubiquitous in the mammalian brain. It is estimated that GABA may function at up to 50% of the synapses and thus it is involved in the central regulation of multiple physiological processes. The GABA$_A$ receptors are the most abundant mediators of GABAergic neurotransmission. They are GABA-gated chloride ion channels and they are now known to be members of the ligand-gated ion channel super-family. In addition, the GABA$_A$ receptors are the site of action for several classes of centrally-acting drugs which include the benzodiazepines, the barbiturates, some steroids and non-competitive antagonists of GABAergic neurotransmission such as the cage convulsant compounds and picrotoxin (1). Indeed, it was the co-existence of the binding site for the benzodiazepines on the GABA$_A$ receptor oligomers which led to the isolation of this protein to apparent homogeneity by benzodiazepine affinity chromatography. The principal GABA$_A$ receptor was shown to consist of two polypeptide chains, α with M_r 53 000 and β with M_r 57 000 (reviewed in ref. 2). Amino acid microsequencing of the respective subunits led to the molecular cloning of the α and β subunits and when the respective RNAs were co-injected in *Xenopus* oocytes, GABA-gated chloride ion channels were formed (3). This led to the proposed model of the GABA$_A$ receptor as depicted in *Figure 1*. The availability of these new probes with which to study the GABA$_A$ receptors now led to important discoveries. First, cDNA homology hybridization identified isoforms of the α subunit. These were α1 (original), α2, and α3 (4). The synthetic oligomers α1 +β, α2+β and α3+β when expressed in *Xenopus* oocytes showed differential sensitivity to agonist activation (4). More recently, additional isoforms of the α subunit α4, α5, and α6, the β subunit β2, β3 and β4 and three new types of GABA$_A$ receptor polypeptides, γ1 and γ2, δ and Σ have been described (5–7). Interestingly, it was shown that α, β, or

Molecular approaches to structure and function of GABA$_A$ receptors

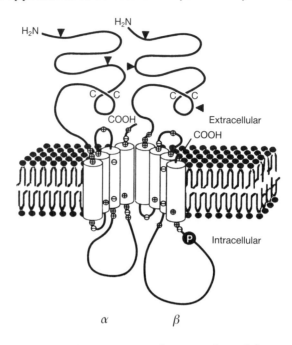

Figure 1. A proposed model for the transmembrane topology of the mammalian GABA$_A$ receptor subunits. The diagram shows the predicted transmembrane organization for the α and the ß subunits and their isoforms. The γ2 polypeptide has the same topology. The subunit stoichiometry and the ordering of the respective subunits are not known. All the polypeptides are predicted to have four membrane-spanning helices which are shown as cylinders; ▼ represents the potential sites for N-glycosylation and Ⓟ is the cyclic AMP-dependent protein kinase site of the ß subunit. A consensus sequence for tyrosine kinase phosphorylation is found in a similar position for the γ2 subunit. The charged residues shown are only those that are located adjacent to the ends of the membrane domains and that are assumed to be involved in chloride ion binding. The inner core of the anion channel is postulated to be formed by four or five copies of the M2 helix one contributed by each polypeptide present in the oligomer. The serine and the threonine residues which are present in the M2 helices are thought to be essential for the flow of the chloride ions. (Reprinted with permission from ref. 3, Copyright © 1987, Macmillan Magazines Ltd.)

γ subunits when expressed singly in *Xenopus* oocytes or in transfected cells formed GABA-gated homo-oligomers (6–9).

Thus it is apparent that the GABA$_A$ receptor system is much more complex than the original protein chemistry studies showed. Indeed, their subunit structure(s) are not known although it is tempting to speculate that multiple iso-oligomers exist and each may have its own pharmacology and specialized function. The approach of molecular cloning and expression of the recombinant receptors provides invaluable information but it does not address directly, what occurs in the central nervous system at the protein level. Our approach is to use the primary structures that have been provided by the molecular cloning studies of the $GABA_A$ receptor polypeptides. We have made antibodies to synthetic peptides the sequences of which are contained within the respective $GABA_A$ receptor subunits and we have used these antibodies to study $GABA_A$ receptor proteins *in vivo*. We describe in this chapter some of our experiences.

2. Synthetic peptides as antigens

Polyclonal and monoclonal antibodies raised against native proteins have been used widely to study both protein distribution and protein structure (for example see ref. 10). The specificity of these antibodies is dictated by the antigenic determinants of the protein of interest and often, the epitopes to which these antibodies bind are formed by amino acids that are not contiguous in the primary structure of the protein, i.e. these antibodies are conformationally-dependent. In contrast, antibodies raised against synthetic peptides have the advantage that they are of pre-determined specificity. Thus synthetic peptides are of particular value for the production of antibodies against receptor or enzyme isoforms for the study of their respective distributions and isoform purification by immunoaffinity chromatography. They can also be employed as probes with which to study receptor transmembrane topology and in conjunction with proteolytic or chemical methods of protein cleavage, they can be used to map the ligand-binding domains within the protein structure. The disadvantage in the use of synthetic peptides as antigens is that they depend upon both the conformation of the chosen peptide being the same as that for the native protein and that the peptide sequence of the protein is accessible to the antibody molecule, i.e. it is on the surface of the protein. These two criteria are not always fulfilled and experience has shown that although anti-peptide antibodies can recognise native proteins, they are more likely to react with denatured proteins as studied by immunoblotting. Thus the search for specific anti-peptide antibodies which recognize native receptors can be both time-consuming and labour intensive.

There are three points to consider in the choice of a peptide sequence for antibody production. These are the actual amino acid sequence itself, the length of the amino acid sequence and if required, the method of covalent coupling of the synthetic peptide to the carrier protein. These points have been addressed in general terms in (11) but we consider them below with specific reference to the $GABA_A$ receptor.

2.1 The choice of synthetic peptide

In the choice of the peptide sequence for antibody production, there are methods available for the prediction of antigenic determinants of proteins.

Molecular approaches to structure and function of GABA$_A$ receptors

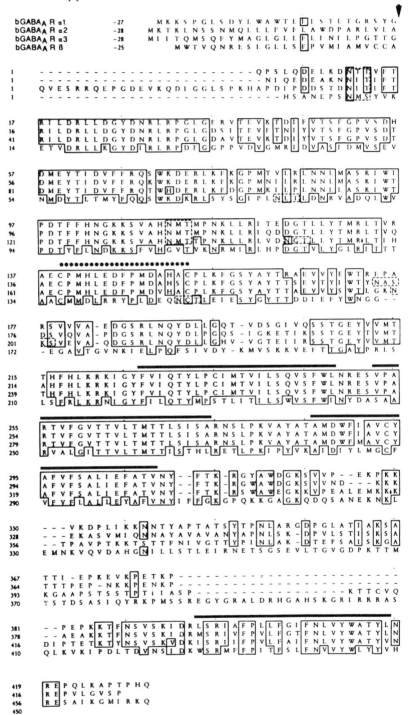

For example, there is the method of Hopp and Woods (12) in which the antigenic determinant is predicted by the local average hydrophilicity of the amino acids. Alternatively, there is the method of Welling and colleagues (13) in which the antigenic regions are predicted by the relative occurrence of each amino acid in known antigenic regions of proteins. However, the choice of the peptide for our purposes was in part dictated by the questions we asked. For example, in the first instance, we wanted to make antibodies specific for each of the isoforms of the $GABA_A$ receptor subunits. A comparison of the amino acid sequences of the α1, α2, and α3 subunits showed that within the trans-membrane regions and the majority of the hydrophilic N-terminal domain, the conservation in amino acid sequence is 80% whereas the most divergent regions of the polypeptides are found at the respective N-termini, the C-termini and within the putative cytoplasmic loop domains (*Figure 2*). The cytoplasmic loop region contains the most divergent sequences between the isoforms and indeed, we have exploited this with the production of antibodies to the sequence α1 324–341. However, it was advantageous also to make antibodies to the N- and the C-termini of the α1 subunit because although the alignment of the sequences showed identities there was no contiguous amino acid sequence identity within these regions. The termini of a protein are often antigenic (14) thus these were good candidate sequences for the production of antibodies that recognize the native receptor and in addition, antibodies to these regions would permit studies on the orientation of the receptor in the membrane. We chose the sequences α1 1–15 and α1 413–429 (15). It should be noted that in the C-terminal sequences the sequence ATYLNR is conserved between isoforms but it is within the fourth transmembrane domain M4. It is hydrophobic in nature and therefore presumably not available for antibody binding. It has been shown that it does not immunocrossreact between isoforms of subunits (15). Some of the $GABA_A$ receptor polypeptides have no C-terminal hydrophilic tail. For example, the β1 and the γ2 subunits where the amino acid sequences terminate immediately at the end of the predicted M4 domain (see later). For these subunits, this region is not a candidate for anti-peptide antibody production.

Figure 2. A comparison of the deduced amino acid sequences of the bovine $GABA_A$ receptor α1, α2, α3, and β1 subunits. This shows an alignment of the amino acid sequences of the respective polypeptides, and identities between them are boxed. The amino acid sequence numbering starts at the predicted N-terminus of the mature polypeptide and ▼ denotes the potential signal sequence cleavage site. — are the proposed membrane-spanning domains M1–M4; .... is the predicted disulphide β loop structure common to all members of the ligand-gated ion channel super-family, and – – – – shows the putative extracellular N-linked glycosylation sites, 2 in α1 subunit; 3 in α2 subunit, and 4 in α3 subunit. (Reprinted with permission from ref. 4, Copyright © 1988, Macmillan Magazines Ltd.)

The length of the synthetic peptide chosen was determined to some extent by the high cost of peptide synthesis. Thus the peptides that we used were a compromise between the minimum length of seven amino acids which has been reported to be required for a satisfactory reaction with intact protein (16) and the length of twenty amino acids that was reported by Lindstrom for the production of antibodies directed against the nicotinic acetylcholine receptor (17). Thus the peptides we used varied from 10–19 amino acids in length. These peptides were either synthesized on a Cambridge Research Biochemicals Pepsynthesiser II using 9-fluorenylmethoxycarbonyl-protected pentafluorophenyl-activated ester amino acids with a Pepsyn KA solid support or they were purchased from the Institute of Animal Physiology and Genetics Research Station, Cambridge, UK, or from Multiple Peptide Systems Inc. San Diego, California, USA. The peptides used were > 85% pure by HPLC analysis.

In all cases, the peptides were coupled to a carrier protein as described below. For the majority of the peptides, keyhole limpet haemocyanin (KLH) was used as the carrier protein. In our initial studies however soybean trypsin inhibitor was employed. We observed strong immunoreactivity of the antibodies against a peptide–soybean trypsin inhibitor conjugate with purified receptor in the ELISA screening assay. We later realized that soybean trypsin inhibitor is used as a protease inhibitor in the isolation of the $GABA_A$ receptor and although the soybean trypsin inhibitor was not detected in SDS-PAGE of the isolated receptor, sufficient soybean trypsin inhibitor was carried through the purification procedure to give a reaction with the antibodies thus occluding the response of the anti-peptide antibodies with receptor. Sato and Neale (18) have reported the production of antibodies to the sequences α1 174–203 and β1 170–199 of the bovine $GABA_A$ receptor. These antibodies recognize the denatured receptor in immunoblots. In this case, the peptide lengths were 30 residues and no carrier protein was employed.

2.2 The method of covalent coupling of peptide and carrier

With regard to the method of coupling of the peptide to the carrier protein, we have used four different methods. These are:

(a) the glutaraldehyde method which links the peptide via the terminal amine group or the ε–NH_2 of lysine to the corresponding site on the carrier protein.

(b) the bis-diazo-*o*-tolidine method which couples tyrosine residues.

(c) the carbodiimide method (EDAC) which couples the peptide via the C-terminal carboxyl group to primary amine sites on the carrier.

(d) The maleimido-benzoic acid *N*-hydroxysuccinimide ester (MBS) method which couples the cysteine of the peptide to the primary amines of the carrier.

These methods are summarized in *Protocols 1–4* with an example of the peptide that was used in each use.

Protocol 1. The glutaraldehyde method for the coupling of peptide and carrier protein via primary amine groups

This method was used for the sequences α1 1–15, QPSLQDELKDNTTVF, α1 324–341, PEKPKKVKDPLIKKNNTY and α1 413–429, ATYLN-REPQLKAPTPHQ.

1. Dissolve keyhole limpet haemocyanin (KLH) in 0.1 M $NaHCO_3$ to a final concentration of 2 mg ml^{-1}.

2. Add glutaraldehyde (Sigma, Grade 1 stored as a 25% (w/v) aqueous solution at −20°C) to a final concentration of 0.05% (v/v). The glutaraldehyde should be taken from a fresh vial.

3. Mix the sample end over end in a glass tube overnight at room temperature.

4. Add 0.1 vol. 1 M glycine ethyl ester (pH 8.0 with NaOH) and mix for 30 min at room temperature.

5. Either (i) precipitate the peptide–KLH conjugate with 5 vol ice cold acetone at −70°C for 30 min. Centrifuge at 10 000 g for 10 min, decant supernatant and dry the pellet *in vacuo*. Resuspend the pellet in 0.9% (w/v) NaCl, at 1 mg KLH ml^{-1} using a Dounce homogenizer
   or (ii) dialyse the peptide–KLH conjugate into 0.9% (w/v) NaCl and adjust the dialysed sample to 1 mg KLH and peptide ml^{-1}.

Protocol 2. The bis-diazo-*o*-tolidine method for the coupling of peptide and carrier protein via tyrosine residues

This method was used for the sequence α1 324–341, sequence as in *Table 1*.

1. Dissolve 0.23 g *o*-tolidine hydrochloride in 45 ml 0.3 M HCl at 4°C. Add 5 ml 0.5 M $NaNO_2$ and stir at 4°C for 60 min. (Aliquots of this, the bis-diazotized *o*-tolidine can be kept at −70°C if rapidly frozen in a cardice bath or liquid nitrogen.)

2. Mix 0.75 ml KLH solution (i.e. 20 mg KLH ml^{-1} in 150 mM NaCl, 160 mM sodium borate, pH 9.0) with 1 ml peptide solution (i.e. 5 mg peptide ml^{-1} in borate buffered saline, pH 9.0 as above). Add 0.6 ml bis-diazo-*o*-tolidine (prepared as in step 1) and adjust the pH to 7–8 with NaOH or boric acid.

3. Incubate on ice for 2 h in the dark with occasional shaking. A fine dark-red precipitate will form.

4. Dialyse against 0.9% (w/v) NaCl (1 litre) overnight at 4°C. The peptide–KLH conjugate (1 mg peptide ml^{-1}) is now ready to use. Do not precipitate because the precipitated conjugate is nearly insoluble and extremely difficult to redisperse adequately.

Protocol 3. The 1-ethyl-3-(3-dimethylaminopropyl)-carbodiimide (EDAC) method for the coupling of peptides and carrier protein via the —CO$_2$H of the peptide to the primary amine of the carrier

1. Dissolve 5 mg peptide in 1 ml 1 mM HCl at 4°C.
2. Add 5 mg EDAC and vortex until dissolved. Leave to stand on ice, with occasional mixing for 20 min.
3. Add 1 ml KLH (20 mg ml^{-1} in 0.5M NH$_4$CO$_3$, pH 9.0) and mix overnight at room temperature.
4. Dialyse exhaustively against 0.9% (w/v) NaCl (2 × 1 litres) and adjust to a final concentration of 1 mg peptide ml^{-1}.

Protocol 4. The *m*-maleimidobenzoic acid *N*-hydroxysuccinimide ester (MBS) method for the coupling of peptides and carrier protein via the cysteine of the peptide to the primary amine of the carrier protein

This method was used for the sequences α1 139–153, CPMHLEDFPM-DAHACG and Cys α3 479–492, CVNRESAIKGMIRKQ.

1. Dissolve KLH at 20 mg ml^{-1} in 10 mM potassium phosphate, pH 7.2. Dialyse overnight at 4°C against 10 mM phosphate buffer pH 7.2 and adjust the concentration to 16 mg KLH ml^{-1}.
2. The presence of a cysteine residue within the peptide gives the possibility of oxidation. If it is suspected that this is the case or if the peptide has been stored as a stock solution, it should be reduced before use. To reduce, dissolve the peptide at 10 mg ml^{-1} in potassium phosphate pH 7.2. Add solid dithiothreitol to a final concentration of 200 mM; dissolve and incubate for 1 h at room temperature. Desalt the peptide by gel filtration column chromatography in 10 mM potassium phosphate pH 7.2. For the desalting, we have used Bio-Gel P-2 which has a molecular weight cut-off of 1800 daltons and it is from Bio-Rad Laboratories, California, USA.
3. Take 250 μl of dialysed KLH and add slowly 85 μl MBS with mixing. The stock MBS solution is 3 mg ml^{-1} in dimethylformamide. Mix end over for 30 min at room temperature.

4. Desalt the activated KLH on a 20-ml Bio-Gel P-30 column equilibrated with 50 mM potassium phosphate pH 6.0. Collect 1 ml fractions. The activated KLH is eluted in the exclusion volume of the column in fractions 6–8. The recovery of KLH is ~ 95%.
5. Dissolve the peptide in 10 mM potassium phosphate pH 7.2 to a final concentration 5 mg ml^{-1}. Add 1 ml of this solution to the activated-KLH (3 ml) mix and adjust to pH 7.4. Mix for 3 h at room temperature.
6. Add solid NaCl to a final concentration of 0.9% (w/v). The peptide-conjugate now at a concentration of 1 mg peptide ml^{-1} is ready for use.

The choice of the coupling method was governed by the peptide sequence. Thus it was preferred that the peptide was linked to the carrier protein via the N- or the C-terminal amino acid since in this case the peptide would be most able to adopt its natural conformation in solution. With the methods that are used for terminal amino acid coupling, they will also react with any lysines in the internal sequence (the glutaraldehyde method) or aspartate or glutamate residues (the EDAC method). This may be a disadvantage because coupling at an internal site may restrict the conformation of the peptide. However, it should be noted that with the sequence α1 324–341, there are six internal lysines but no difference was found between the reactivity of the antibodies produced for either glutaraldehyde or tyrosine C-terminal coupling to the carrier protein. In some cases, it was thought appropriate to add an N- or C-terminal cysteine which was not contained within the natural sequence, e.g. Cys-α3 479–492. Here there was the advantage that the peptide bond is now present at the N-terminal end of the peptide thus more closely resembling the natural protein than if it was just the free amino group. In some cases where an internal sequence of a $GABA_A$ receptor subunit was taken, the C-terminal amino acid was amidated again to resemble the natural protein. Sometimes, it was fortuitous or by design that the end amino acids were either a natural occurring cysteine, e.g. α1 139–153, or tyrosine, e.g. α1 324–341, which were then used for the coupling of the peptide to the carrier protein as described in *Protocols 4* and *2* respectively.

3. Antibody production and antibody screening methods

For each peptide-carrier conjugate, polyclonal antibodies were raised in Dutch-belted rabbits. The immunization regimen that was followed was that the respective peptide-carrier conjugates (0.2 μmol peptide) were emulsified with an equal volume of Freund's complete adjuvant and the primary injection was at two sites intramuscularly. Subsequent injections were in Freund's incomplete adjuvant again with 0.2 μmol peptide at two sites intramuscularly. The animals were ear bled at seven days following the second and subsequent immunisations. For each peptide-carrier conjugate, two rabbits were employed and to date, the antibody response has been the same for each animal. That is that the anti-peptide antibodies produced in both rabbits either recognize the $GABA_A$ receptor or they fail to react with the receptor in any of screening methods used.

Three different methods have been employed for the screening of the immune sera. These are:

- a solid-phase enzyme-linked immunoabsorbent assay (ELISA) with either the respective peptide or the $GABA_A$ receptor as antigen;
- screening by immunoblotting and
- screening by a soluble immunoprecipitation assay.

3.1 The ELISA and immunoblotting screening methods

In each case the initial screening method of the immune sera was the ELISA because it was already used routinely for the screening of animals immunized with receptor (19) and alone of the above methods, it was easily adapted to measure the production of anti-peptide antibodies themselves. The ELISA methods are summarized in *Protocol 5*. To date, from fourteen different peptides whose sequences are derived from $GABA_A$ receptor subunit primary structures only one of the peptides α1 210–223 has failed to generate an immune response which was readily detected by the appropriate ELISA. In general, high titres of anti-peptide antibodies were produced. This was in contrast to the antibody titre for the reaction of the antibodies with purified $GABA_A$ receptor which was always lower and this was expected since the peptide-carrier conjugate can be assumed to adopt multiple conformations in solution and not all of these will be the same as that for the $GABA_A$ receptor.

Protocol 5. Enzyme-linked immunoabsorbent assay (ELISA) for the measurement of anti-peptide and anti-$GABA_A$ receptor antibodies

1. Coat the wells of the plate with antigen. The plates we use are polyvinyl Microtest III flexible assay plates from Becton Dickinson and Co., Oxnard, California, USA.

    (a) *Synthetic peptides as antigens*. The optimum conditions will vary between peptides, but satisfactory results can be obtained by incubating the peptide in the well at 1 μg ml^{-1} in 0.05 M $NaHCO_3$, 0.01M NaOH pH 9.5 overnight at 4°C (100 μl per well).

    (b) *$GABA_A$ receptor as antigen*. Dilute the $GABA_A$ receptor purified by benzodiazepine affinity chromatography as described by Sigel *et al.* (1983) except that the detergent concentration was reduced to 0.05% Triton X-100 in the ion-exchange step with 15 vol of phosphate buffered saline, pH 7.4. Pipette 100 µl into each well and leave at least overnight at 4°C. Improved efficiency of coating can be achieved by incubation for 2–3 days at 4°C.

2. Wash wells three times with 200 µl of PBS containing 0.25% (w/v) gelatine (PBS-gelatine). Add a further 200 µl PBS-gelatine and incubate for 45 min at 37°C.

3. Aspirate and discard PBS-gelatine and add 100 µl of the test serum diluted in PBS-gelatine. We normally assay a highest concentration of 1:10 diluted sera and thereafter serial dilutions every 0.5 $\log_{10}$ units. Eight wells are used for each serum. Incubate for 1 h at 37°C.

4. Aspirate serum samples and wash the wells three times with PBS-gelatine. Add 200 µl PBS-gelatine and incubate for 10 min at 37°C.

5. Aspirate the PBS-gelatine and add 100 µl of biotinylated anti-rabbit Ig (Amersham International plc, UK) diluted 1:750 in PBS-gelatine. Incubate for 1.5 h at 37°C.

6. Remove the second antibody and wash the wells four times with 200 µl PBS-gelatine.

7. Remove PBS-gelatine and add 100 µl pre-formed streptavidin–horseradish peroxidase complex (Amersham International plc) diluted 1:1000 in PBS-gelatine. Incubate for 45 min at 37°C.

8. Aspirate the horseradish peroxidase and wash the wells three times with 200 µl PBS-gelatine and once with 200 µl PBS.

9. Immediately prior to use prepare the substrate which is 4 mM o-phenylenediamine in 0.02 M citric acid and 0.25 M Na_2HPO_4 pH 5. Add H_2O_2 to a final concentration 0.004% (v/v). Add 100 µl of the substrate mix to each well and leave to develop. Stop the reaction by adding 50 µl 20% H_2SO_4 to each well.

10. Read the optical density at $\lambda = 492$ nm.

The development of the immunoreactivity of the anti-peptide antibody with the receptor varied between the peptides used for immunization. For example, anti-α1 1–15 antibodies reacted with $GABA_A$ receptor after the first test bleed whereas anti-α1 413–429 antibodies reacted with receptor only after the third test bleed. But for each, once a reaction with receptor was obtained the titre increased with subsequent immunizations to a plateau level. A summary of the antibody titres and their development for a selection of the peptides employed is given in *Table 1*.

In the ELISA assay, the receptor in its native state is coated on to the walls of the plastic wells. The nature of the interaction between the receptor and the plastic is not known but it has been shown that it must retain some of its

Table 1. A summary of the antibody titres obtained for rabbits immunized with synthetic peptide derived from GABA$_A$ receptor subunit sequences

Peptide	1* peptide+	1* receptor'	2* peptide+	2* receptor'	3* peptide+	3* receptor'	4* peptide+	4* receptor'	5* peptide+	5* receptor'	6* peptide+	6* receptor'
α1 1-15	200	100	1000	150	3000	825	3000	750	3500	750	3500	1000
	200	100	1000	100	2000	400	2500	400	2500	400	4500	800
α1 324-341	1000	0	2000	0	2000	0	5000	150	6000	500	5000	200
	1500	0	2500	0	2600	0	5000	150	5000	200	3000	65
	37000	2400	20000	1000	15000	NT	NT	NT	NT	NT	NT	NT
α1 413-429	200	0	500	0	1000	50	2000	800	1500	400	NT	NT
	300	0	600	0	1000	0	1200	250	1200	100	NT	NT
α2 323-339	190	0	480	0	260	0	130	0	100	0	70	0
	140	0	160	0	200	0	190	0	90	0	50	0
Cys-α3 479-492	0	0	50000	0	10000	30	5000	80	3000	100	3000	100
	9000	0	90000	0	30000	30	40000	80	30000	100	20000	100

* 1–6 refers to the bleed number of the rabbits where the first bleed was seven days following the second immunization.
NT = not tested, and + and ' refer to the antibody titre obtained in the ELISA against the respective peptide or GABA$_A$ receptor purified from adult bovine cerebral cortex as antigen. The antibody titre is expressed as the fold-dilution of immune serum which gave 50% of the maximum absorbance at λ = 492 nm. for each peptide, the results for each animal are presented.

native conformation because [^{3}H] flunitrazepam will bind specifically to this immobilized receptor (S.O. Casalotti and F. A. Stephenson, unpublished observations). If the anti-peptide antibody is negative in the ELISA assay with receptor as antigen this may be therefore, because the specific antibody will recognize denatured receptor only. Immunoblots which use purified denatured receptor as antigen are always used as the second method of screening and this is described in *Protocol 6*.

Protocol 6. Immunoblots using anti-peptide antibodies against the GABA$_A$ receptor

1. Run the antigen, i.e. the purified receptor or brain membranes in a 10% polyacrylamide slab gel under reducing conditions as for standard methods of SDS-PAGE.
2. Transfer the antigens on to a nitrocellulose membrane filter which has been pre-soaked in the transfer buffer which is 25 mM Tris 192 mM glycine 20% (v/v) methanol pH 8. Transfer the antigens to the membrane filter by electrophoresis at 30 V constant voltage overnight and 1 h at 50 V, all at 4°C.
3. Block the non-specific protein binding sites by incubation with 50 mM Tris, 0.9% (w/v) NaCl pH 7.4 (Tris-saline) containing 5% (w/v) dried skimmed milk powder and 0.1% (v/v) Tween-20, for 1 h at 37°C.
4. Incubate the nitrocellulose filter with immune serum (typically a 1:50 dilution in Tris-saline containing 2.5% (w/v) dried skimmed milk) for 1 h at 37°C.
5. Wash four times, 10 min at 37°C each with Tris-saline containing 2.5% (w/v) dried skimmed milk) powder and 0.1% (v/v) Tween.
6. Incubate the nitrocellulose membrane with biotinylated anti-rabbit Ig (Amersham International plc), diluted 1:500 in Tris-saline containing 2.5% (w/v) milk powder.
7. Wash in Tris-saline containing 2.5% (w/v) milk powder at 0.1% (v/v) Tween four times for 10 min each at 37°C.
8. Incubate with pre-formed horseradish peroxidase–streptavidin complex (Amersham International plc, diluted 1:400 with Tris-saline containing 2.5% (w/v) milk) for 20 min at room temperature.
9. Develop by incubation with peroxidase substrate. The substrate mix is prepared immediately prior to use and is 0.5 mg ml^{-1} diaminobenzidine, 0.3 mg ml^{-1} CoCl$_2$, 0.02% (v/v) H$_2$O$_2$.
10. Stop the reaction when the colour has developed by exhaustively washing in distilled water.

An alternative method that was used for the detection of immunoreactive

Molecular approaches to structure and function of $GABA_A$ receptors

proteins was the use of ^{125}I protein A as demonstrated in *Figure 3*. In this case, the immunoblotting was carried out as far as (5) above except that the final wash did not contain Tween. The nitrocellulose membrane was

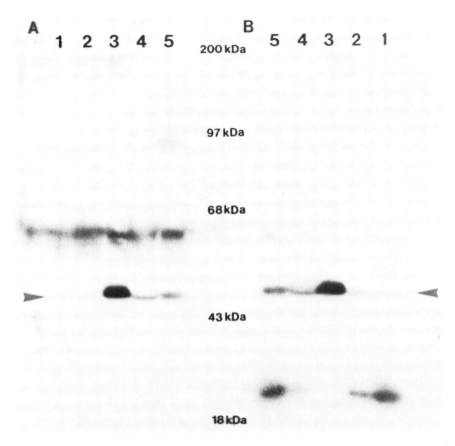

Figure 3. Brain-regional expression of the α1 subunit of the bovine $GABA_A$ receptor. This shows an immunoblot of membranes prepared from different adult bovine brain regions subjected to sodium dodecyl sulphate polyacrylamide gel electrophoresis in 10% slab gels under reducing conditions and then probed with anti-α1 1–15 and anti-α1 413–429 antibodies by Western blotting as described in *Protocol 7* with the visualization of immunoreactive bands with ^{125}I protein A. The blot is representative of three experiments and in each case, 0.5 pmol [^{3}H] flunitrazepam binding sites were applied for each membrane sample which corresponded to 0.54, 0.6, 0.23, 0.5 mg protein for cerebral cortex, cerebellum, hippocampus, and striatum respectively. **A** is blotting with anti-peptide α1 1–15 antibodies and **B** is blotting with anti-peptide α1 413–429 antibodies. **Lanes** are **1**, striatum; **2**, hippocampus; **3**, $GABA_A$ receptor purified from adult bovine cerebral cortex as positive control; **4**, cerebellum and **5**, cerebral cortex. The numbers show the positions of molecular weight standards and ◄ show the position of the α1 subunit. (Reprinted from ref. 15 with the kind permission of Raven Press.)

incubated with ^{125}I protein (2–5 × 10^5 c.p.m./ml) in Tris-saline containing 5% (w/v) dried skimmed milk for 40 min at room temperature. It was washed at room temperature, 15 min each with three times Tris-saline containing 5% (w/v) dried skimmed milk, 0.1% (v/v) Tween, once with Tris-saline containing 5% (w/v) dried skimmed milk and once with Tris-saline. It was dried by blotting with Whatmann 3 mm paper and exposed to pre-flashed Fuji X-ray film at room temperature for 1–2 weeks.

In both the ELISA and the immunoblotting screening methods, it is important to be aware of the fact that the sensitivity is dependent upon the nature of the antigen. This becomes important in particular in the study of the distribution of the isoforms of the GABA$_A$ receptor polypeptides which may themselves each have a brain-regional distribution. Northern blots have indeed shown that the most abundant transcript of the α subunits in all brain regions is the α1 subunit mRNA whereas the α2 mRNA is the least abundant although it is enriched in the hippocampus (4). The antigen we use routinely is the GABA$_A$ receptor purified from adult bovine cerebral cortex which may not therefore contain sufficient α2 subunit for the detection of anti-α2 subunit immunoreactivity; GABA$_A$ receptor purified from hippocampus would be preferable for use as an antigen in this case.

In accordance with the results of the Northern blots, Western blots, and immunoprecipitation studies (see below) using anti-peptide antibodies have demonstrated that, at the protein level, the most abundant of the α subunit isoforms is the α1 subunit. Indeed the α1 polypeptide as well as α1 mRNA has a brain-regional distribution (15). *Figure 3* shows the results obtained when immunoblots were carried out with anti-α1 1–15 and anti-α1 413–429 antibodies against membranes prepared from different brain regions. The immunoblot shown in *Figure 3* was carried out using diluted immune serum and it is notable that other polypeptides are recognized by the antibodies. We now use affinity-purified antibodies for the immunoblotting and immunoprecipitation experiments which yield cleaner results. A general method for the affinity purification of the anti-peptide antibodies is given in *Protocol 7*

Protocol 7. Purification of anti-peptide antibodies by peptide affinity chromatography

There are two stages in the affinity purification of the anti-peptide antibodies. These are, first, the construction of the appropriate immobilized peptide affinity column, and, second, the use of this affinity column for the purification of the anti-peptide antibodies. Examples are given below for the construction of peptide affinity columns via primary amine or sulphydryl groups.

Preparation of peptide affinity resins
(a) Coupling via primary amine groups

1. Swell 0.35 g Activated CH-Sepharose 4B (Pharmacia) in H_2O and wash with 100 ml 1 mM HCl at 4°C on a sintered glass filter. Equilibrate gel with 0.1 M $NaHCO_3$ pH 8.0 containing 0.3 M NaCl and transfer to a capped tube in 1 ml of the equilibration buffer.

2. Add 1 ml of 5 mg ml^{-1} peptide in the equilibration buffer and mix end over end for 1 h at room temperature. We have used the peptide α1 324–341).

3. Terminate the reaction by washing the gel with equilibration buffer (25 ml) on a sintered glass filter. Transfer the gel to a capped tube and block all the remaining active groups by resuspension in 0.1 M Tris–HCl pH 8.0 containing 0.5 M NaCl (3 ml) and mixing end over end for 1 h at room temperature.

4. Wash the gel four times each and alternately with 0.1 M CH_3COOH adjusted to pH 4.0 with NaOH containing 0.5 M NaCl and 0.1 M Tris–HCl pH 8.0 containing 0.5 M NaCl.

5. Pour the column and equilibrate with phosphate buffered saline and use immediately or store in the presence of 0.02% (w/v) NaN_3 at 4°C.

This procedure is for a 1-ml affinity column and it has been employed to purify anti-peptide antibodies from at least 5 ml immune serum.

(b) Coupling via sulphydryl groups

1. Swell 0.35 g Activated Thiol–Sepharose 4B (Pharmacia) in H_2O and wash on sintered glass filter with 100 ml 0.1 M Tris HCl pH 8.0 containing 0.3 M NaCl, 1 mM EDTA which has been degassed under vacuum. Transfer to a capped tube.

2. Add 1 ml of 5 mg ml^{-1} peptide in the wash buffer as above in step 1 and mix end end over for 2 h at room temperature. (We have used the peptide Cys-α3 479–492.)

3. Terminate the reaction by washing the gel on a sintered glass filter with the Tris equilibration buffer (25 ml) then wash with 100 mM citric acid adjusted to pH 4.5 with 2 M KOH (10 ml). Block the unreacted thiol groups by incubating the gel with 1 mM β-mercaptoethanol in citrate buffer (3 ml) and end over end mixing for 45 min at room temperature.

4. Terminate the blocking reaction by washing with 100 mM citric acid pH 4.5 (25 ml).

5. Pour the column and equilibrate with phosphate buffered saline and use immediately or store in the presence of 0.02% (w/v) NaN_3 at 4°C.

This procedure is for a 1 ml affinity column and it has been employed to purify anti-peptide antibodies from at least 5 ml immune serum.

Affinity purification of anti-peptide antibodies

1. Equilibrate the appropriate peptide affinity resin (1 ml) with phosphate buffered saline. Add the respective immune serum, (1–5 ml) and recirculate through the column at a rate of 40 ml h^{-1} for 2 h at room temperature or overnight at 4°C.
2. Collect the filtrate and wash the column with 100 ml phosphate buffered saline at 40 ml h^{-1}.
3. Elute the antibody with 10 ml 50 mM glycine/HCl pH 2.3 at 10 ml h^{-1} and collecting 1 ml fractions. Neutralize each fraction immediately after elution by the addition of 1 M Tris (20 μl) to give a final pH of 7.4.
4. The efficiency of the purification procedure was monitored by ELISA on all appropriate fractions. The yield of the purified antibody was in the range of 0.2–0.6 mg protein ml^{-1}.

The use of anti-peptide antibodies in conjunction with immunoblotting enabled us to identify the α3 isoform of the subunit of the GABA$_A$ receptor (20). The α3 subunit was discovered by cDNA homology cross-hybridization with an α subunit cDNA probe (4). It was predicted to have a molecular weight M_r 52 000 daltons compared to the predicted size of the α1 subunit of M_r 48 000 daltons. The α3 subunit contains consensus sequences for N-glycosylation in predicted extracellular domains. We observed that the size of the α3 subunit was 59–60 000 daltons which is compatible with a glycosylated α3 polypeptide with a contribution of 7–8000 daltons by weight by the carbohydrate (20). Kirkness and Turner (21) made antibodies to the sequence α1 101–109. This amino acid sequence is conserved between the α1 and the α3 subunit isoforms and indeed they found the recognition of two bands with M_r 52000 and M_r 58–59 000 daltons by their antibody in Western blots. Their findings highlight the caution needed in the choice of an amino acid sequence for antibody production for specific isoforms of subunits (see earlier).

3.2 The soluble immunoprecipitation assay

The third method that has been used to characterize the anti-peptide antibodies is that of immunoprecipitation. Of the three methods that have been employed, immunoprecipitation alone requires that the antibody recognizes the native receptor in detergent solution. It is essential that the antibody will specifically immunoprecipitate receptor if it is to be further employed for immunoaffinity purification or immunochemical tissue distribution studies of receptor isoforms. The method that we have used routinely is outlined in *Protocol 8*. It is a standard soluble assay based on the reaction between antigen and antibody, precipitation of the antigen–antibody complex by either second antibody or protein A in the form of Immunoprecipitin

(Bethesda Research Laboratories, Bethesda, Maryland, USA) which is formalin-fixed *Staphylococcus aureus*. The ability of the antibody to recognize antigen is then monitored by either a decrease in binding activity in the supernatants or an increase in ligand-binding activity in the pellets following centrifugation. The antigen can be either:

- purified $GABA_A$ receptor, detergent-solubilized receptor, or reconstituted receptor, and for these cases reversible ligand-binding assays to the supernatants and pellets are carried out, *or*
- [^{3}H] flunitrazepam or [^{3}H] muscimol photoaffinity labelled receptor and in these cases the pellets and supernatants can be counted directly (*Protocol 8*).

Protocol 8. Immunoprecipitation of the purified and detergent-solubilized $GABA_A$ recptor from adult bovine cerebral cortex

1. Incubate the antigen and antibody for 3 h at room temperature or overnight at 4°C in a final volume of up to 500 µl. The volumes of the antigen and the antibody used will vary but we have used:

    - Purified $GABA_A$ receptor: 50 µl of the pool of the ion-exchange chromatography fractions as in ref. 22.
    - Soluble $GABA_A$ receptor: 200 µl of the sodium-deoxycholate solubilized receptor prepared as in ref. 22.
    - Immune sera: up to 40 µl.
    - Affinity purified antibody: up to 200 µl, with a typical protein concentration up to 250 µg/ml.

    In each case, the final assay volume was made up by the addition of 10 mM potassium phosphate pH 7.4 containing 150 mM KCl. It is essential that all tubes contain an equivalent quantity of immunoglobulin and this is done by either diluting the immune serum with normal rabbit serum (pre-immune serum) or by the inclusion of a control sample which is an equivalent protein concentration of protein A purified non-immune IgG for affinity purified antibody. A control which contains no antibody should also be used to test for non-specific precipitation of $GABA_A$ receptor.

2. Add Immunoprecipitin (formalin-fixed *Staphylococus aureus*, 100 µl) and incubate for 1 h at room temperature.

3. Add 400 µl 10 mM potassium phosphate, pH 7.4 containing 150 mM KCl and 0.5% (v/v) Triton X-100. Vortex, and centrifuge for 15 min at 10 000 g.

4. From the resultant supernatant, take 6 × 150 µl aliquots for the determination of total (3 × 150 µl) and non-specific (3 × 150 µl) ligand-binding activities.

5. Resuspend the pellet in 100 µl 10 mM potassium phosphate, pH 7.4, containing 150 mM KCl and 0.5% (v/v) Triton X-100 by drawing through an air displacement pipette tip several times. Add a further 900 µl of this buffer and centrifuge for 10 min at 10 000 g. Discard the supernatant and repeat the wash procedure twice more.

6. Resuspend the pellet as in step 5 to a final volume of 1 ml. Take 6 × 150 µl aliquots for the measurement of total (3 × 150 µl) and non-specific (3 × 150 µl) ligand-binding activities. Radioligand-binding was measured by the polyethyleneimine method as described in ref. 22.

We found that antibodies raised to the N- and C-termini of the α1 subunit, anti-α1 1–15 and anti-α1 413–429 antibodies both recognized the native protein. Both anti-peptide α1 1–15 and anti-peptide α1 413–429 antibodies immunoprecipitated in parallel [^{3}H] flunitrazepam and [^{3}H] muscimol specific binding activities in a dose-dependent manner from either purified or crude soluble extracts of bovine cerebral cortical GABA$_A$ receptors (15). The maximum number of sites immunoprecipitated, however, plateaued at 55–60% for both ligands in both purified and soluble preparations of cerebral cortex. These results are shown in *Figure 4* and they were interpreted as evidence for the existence of receptor iso-oligomers, i.e. α1 + β1, α2 + β1, etc., rather than the presence of α1 + α2 + α3 + β1 in the same oligomeric structure. They agree with the results of immunoblotting and Northern blots that the α1 subunit is the most abundant of the α subunit isoforms in cerebral cortex. More recently, we have extended our studies to a comparison of the brain and regional distribution of the α1, α2, and α3 subunit isoforms (23).

The GABA$_A$ receptor resembles most proteins whose epitopes have been mapped in that the termini of the protein are antigenic (14). The receptor also has the advantage that for antibodies specific for isoforms of subunits, these domains within the receptor polypeptide chains are divergent. Thus these sequences are obvious candidates for antibody production to the epitopes unique to each of the GABA$_A$ receptor polypeptides and their respective isoforms. Indeed, we predicted that antibodies directed against the sequence cys α3 479–492 (the C-terminus) would recognize the native receptor and this was proved to be correct (20). It is our general experience that antibodies directed at the C-terminal sequences are preferred to those of the N-termini but as mentioned earlier, all the β-sequences and the γ2 subunit sequences terminate immediately after M4 thus excluding the C-terminal sequences from use in these cases.

Molecular approaches to structure and function of GABA$_A$ receptors

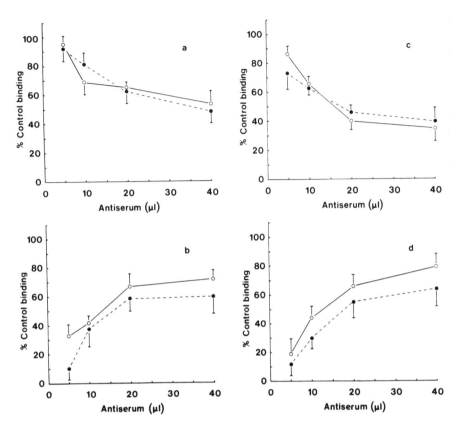

Figure 4. Immunoprecipitation of GABA$_A$ receptor from sodium deoxycholate extracts of adult bovine cerebral cortex with anti-peptide α1 1–15 and anti-peptide α1 413–429 antibodies. Immunoprecipitation was carried out as described in *Protocol 8* using sodium deoxycholate extracts from adult bovine cerebral cortex as antigen and using anti-α1 1–15 (**a** and **b**) and anti-α1 413–429 (**c** and **d**) antibodies. **a** and **c** are the respective ligand-binding activities to the pellets after immunoprecipitation. (○) is specific [^{3}H] muscimol binding activity and (●) is specific [^{3}H] flunitrazepam binding activity. The figure shows a dose-dependent immunoprecipitation of GABA and benzodiazepine binding sites in parallel which plateau at 60 ± 10% ($n = 3$) and 65 ± 9% ($n = 3$) for benzodiazepine and GABA binding sites respectively, for anti-α1 413–429 antibodies and 52 ± 8% ($n = 3$) and 47 ± 12% ($n = 3$) for both sites with anti-α1 1–15 antibodies. (Reprinted from ref. 15 with the kind permission of Raven Press.)

4. Conclusions

In this chapter, we have described the production and characterization of antibodies directed against synthetic peptides derived from GABA$_A$ receptor subunit amino acid sequences. We have focused particularly on the production of antibodies to unique regions of the various $GABA_A$ receptor polypeptides and their isoforms and we have shown how the use of these highly specific probes demonstrated the brain-regional expression of the α1 subunit and identified the novel α3 subunit. The potential use of these antibodies is of course much wider than the examples that we have given. As discussed earlier, they can be employed to determine the various ligand binding domains within the receptor structures and they can be used as probes to test the models of the receptor transmembrane topography as indeed has been done for the nicotinic acetylcholine receptor (17). These studies are not yet completed and therefore beyond the scope of this chapter.

Acknowledgements

The work described was supported by the Medical Research Council (UK). FAS is a Royal Society University Research Fellow. We are grateful to Gill Patterson for secretarial assistance.

References

1. Olsen, R. W. and Venter, J. C. (ed.) (1986). *Benzodiazepine GABA Receptors and Chloride Channels: Structural and Functional Properties*. Alan R. Liss, New York.
2. Stephenson, F. A. (1988). *Biochemical Journal*, **249**, 21.
3. Schofield, P. R., Darlison, M. G., Fujita, N., Burt, D. R., Stephenson, F. A., Rodriguez, H., Rhee, L. M., Ramachandran, J., Reale, V., Glencorse, T. A. Seeburg, P. H., and Barnard, E. A. (1987). *Nature*, **328**, 221.
4. Levitan, E. S., Schofield, P. R., Burt, D. R., Rhee, L. M., Wisden, W., Kohler, M., Fujita, N., Rodriguez, H., Stephenson, F. A., Darlison, M. G., Barnard, E. A., and Seeburg, P. H. (1988). *Nature*, **335**, 76.
5. Pritchett, D. B., Schofield, P. R., Sontheimer, H., Ymer, S., Kettenmann, H. and Seeburg, P. H. (1988). *Society for Neuroscience Abstracts*, **14**, 641.
6. Pritchett, D. B., Sontheimer, H., Shivers, B. D., Ymer, S., Kettenmann, H., Schofield, P. R., and Seeburg, P. H. (1989). *Nature*, **338**, 582.
7. Ymer, S., Shofield, P. R., Draguhn, A., Werner, P., Kohler, M., and Seeburg, P. H. (1989). *EMBO Journal*, **8**, 1665.
8. Blair, L. A. C., Levitan, E. S., Marshall, J., Dionne, V. E., and Barnard, E. A. (1988). *Science*, **242**, 577.
9. Pritchett, D. B., Sontheimer, H., Gorman, C. M., Kettenmann, H., Seeburg, P. H., and Schofield, P. R. (1988). *Science*, **242**, 1306.
10. Venter, J. C., Fraser, C. M., and Lindstrom, J. (ed.) (1985). *Monoclonal and Antiidiotypic Antibodies as Probes for Receptor Structure and Function*. Alan R. Liss, New York.
11. Scheidtmann, K. H. (1989). In *Protein Structure: A Practical Approach* (ed. T. E. Creighton), pp. 93–114. IRL Press at Oxford University Press, Oxford.

12. Hopp, T. P. and Woods, K. R. (1981). *Proceedings of the National Academy of Sciences of the USA*, **78**, 3824.
13. Welling, G. W., Weijer, W. J., van der Zee, R., and Welling-Wester, S. (1985). *FEBS Letters*, **188**, 215.
14. Van Regenmortel, H. V., Altschuh, D., and Klug, A. (1986). In *Synthetic Peptides as Antigens*, Ciba Foundation Symposium 119 (ed. R. Porter and J. Whelan), pp. 76–84. Wiley, Chichester, UK.
15. Duggan, M. J. and Stephenson, F. A. (1989). *Journal of Neurochemistry*, **53**, 132.
16. Welling, G. W. and Fries, H. (1985). *FEBS Letters*, **182**, 81.
17. Lindstrom, J. (1986). *Trends in Neuroscience*, **9**, 401.
18. Sato, T. N. and Neale, J. H. (1989). *Journal of Neurochemistry*, **53**, 1089.
19. Mamalaki, C., Stephenson, F. A. and Barnard, E. A. (1987) *EMBO Journal*, **6**, 561.
20. Stephenson, F. A., Duggan, M. J., and Casalotti, S. O. (1989). *FEBS Letters*, **234**, 358.
21. Kirkness, E. F. and Turner, A. J. (1988). *Biochemical Journal*, **256**, 291.
22. Stephenson, F. A. (1990): In *Receptor Biochemistry: A Practical Approach* (ed. E. C. Hulme), p. 177. IRL Press at Oxford University Press.
23. Duggan, M. J. and Stephenson, F. A. (1990). *Journal of Biological Chemistry*, **265**, 3831.

10

In situ hybridization with synthetic DNA probes

WILLIAM WISDEN, BRIAN J. MORRIS, and STEPHEN P. HUNT

1. Introduction: mRNA hybridization using synthetic oligodeoxyribonucleotide probes

In situ hybridization is the colloquial term for a now standard technique used to detect mRNA (e.g. coding for a peptide transmitter or neurotransmitter receptor) in the neuronal or glial cells where it is being synthesized, using a labelled nucleic acid probe of complementary sequence to a portion of the mRNA. The technique has been used to:

(a) Determine the pattern of distribution of particular mRNA molecules within the brain, often in conjunction with immunocytochemistry and ligand binding (1). Unlike the latter two procedures, the labelling obtained with *in situ* hybridization is usually restricted to the cell body [although recently the mRNA encoding the MAP 2 protein has been shown to be localized in dendrites (2)].

(b) Determine the level of gene activity, which can be used as an index of peptide turnover/release in particular cell types (3).

(c) Study changing patterns of gene activity in the nervous system in response to specific stimuli (4), or changes in patterns of gene expression during development (5).

Most techniques in molecular biology or its fringes seem to have a bewildering number of alternative protocols, each with their own proponents, and *in situ* hybridization is no exception. The probes used to hybridize to and detect mRNA can be of four types.

- complementary single-stranded RNA probes (Riboprobes) derived from transcription vector plasmids;
- complementary single-stranded DNA probes derived from M13 vectors;
- double-stranded DNA probes; and
- synthetic complementary oligodeoxyribonucleotides (6).

In situ hybridization with synthetic DNA probes

This chapter will describe the use of synthetic oligonucleotide probes. The method used here was originally developed by Young et al. (7) and Lewis and colleagues (8). We have recently used it with modifications to map the distribution of transcripts coding for the $GABA_A$ receptor in brain (9), and to examine the transriptional activity of various peptide genes in embryonic striatal transplants (10). We have found this method much more convenient and advantageous to use than the other methods because:

- No subcloning is required, which saves time and makes the method more generally accessible to neurobiologists.

- The method presented in this chapter is minimalist in terms of the number of procedures required compared to many other protocols, and the sensitivity of detection appears to be at least comparable, if not greater than that obtained with the other methods.

- Oligonucleotides are very useful when probes are required that cleanly distinguish closely related or alternatively spliced transcripts. It is sometimes difficult to obtain useful restriction fragments for this purpose.

1.1 Methods included in this chapter

Section 2 discusses the construction (i.e. length and sequence) and synthesis of oligonucleotide probes, followed by a procedure to purify them ready for labelling. Section 3 describes the terminal deoxynucleotidyl transferase reaction which is used to label the probes with ^{32}P or ^{35}S. The in situ hybridization protocol is described in Section 4, including the sectioning and storage of tissue. Section 5 refers to control procedures (since the method can produce misleading artifacts), and various parameters of the protocol which can be varied.

The entire procedure is outlined in the form of a flow diagram in Figure 1. In brief, probes are labelled, diluted in hybridization buffer, and applied to sections (cut and stored on a previous occasion). This part of the procedure takes half a day in total. Sections are then hybridized overnight with the probe, and washed the next day in a heated salt solution, followed by dehydration (total time, 2 h), before apposition to X-ray film (days to weeks exposure) or dipping in photographic emulsion (days to weeks exposure). Typical results obtained with X-ray film ar shown in Figure 5. The cellular resolution obtained with photographic emulsion is illustrated in Figure 3.

2. Oligonucleotide probes

2.1 Probe synthesis

We routinely use probes that are 45 nucleotides long (45-mers), consequently the hybridization conditions given below in this chapter are optimized for this length of probe. However, the conditions also apply to other lengths of

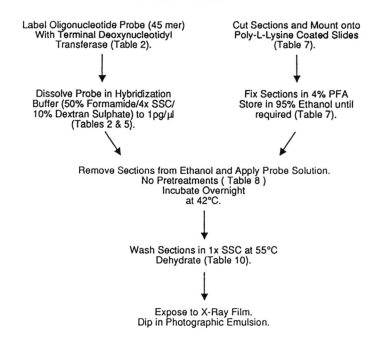

Figure 1. Scheme for *in situ* hybridization with synthetic oligonucleotides. For explanation, see text.

probes we have used, i.e. 42-mers, 48-mers, and 60-mers. Shorter probes require lower incubation temperatures. Access to the services of an oligonucleotide synthesizer (e.g. Applied biosystems or Pharmacia machines) will be required, although some companies will custom build oligonucleotides (e.g. British Bio-technology, Oxford, UK). As many research centres have DNA synthesizers as a central facility, owing to the expense and running costs of the machines, details of oligonucleotide synthesis are not given. However, 1 μg of oligonucleotide will be sufficient for 200 probe labellings.

It is important to remember when designing an oligonucleotide from a published sequence, ideally to have a GC/AT ratio of approximately 50% or more. Oligonucleotides with large numbers of A and T residues will be less stable in forming DNA/RNA hybrids. However, if the GC content is too high (> 65%), this may cause non-specific binding since the thermal stability of the probes will be much greater. If one is preparing a probe to cross between species, it is better to make a longer probe (e.g. as a 60-mer), to compensate for any nucleotide mismatches. Cross hybridization is limited with these probes. An oligonucleotide with say only 90% homology will hybridize poorly, with consequently a substantial loss of signal. Hybridization conditions for oligonucleotides are reviewed elsewhere (11, 12).

In addition to constructing an antisense oligonucleotide, it is also useful to build a probe of the same length from the sense strand for use as a control (see below). Note that the DNA sequences given in papers are usually by convention the sense (coding) strand, so that probes to detect mRNA are built as complimentary sequence to that published, and with their 5' to 3' polarity reversed. For example, if the sense strand sequence is 5'GCAGCCT ...3', then the sequence of the probe will be 3' ... CGTCGGA 5'. This is usually written the other way round as 5'AGGCTGG ... 3'.

2.2 Purification of probes

Once the oligonucleotide has been synthesized and deprotected, it is advisable to purify it further, since the labelling enzymes may be sensitive to contaminating salts from the synthesis. The method given in *Protocol 1* can be used for this. To check that the probes are the correct length they should be kinased and run on a 10% acrylamide/8 M urea gel using other kinased oligonucleotides as markers (see below).

Protocol 1. SEP-PAK purification of synthetic oligonucleotides

1. Prepare the cartridge[a] by passing through 10 ml HPLC grade acetonitrile, followed by 10 ml DEPC treated water.
2. Prepare 10 ml of methanol (HPLC grade)/sterile water (60:40).
3. Dissolve the oligonucleotide[b] in 2 ml of sterile 0.5 M ammonium acetate, 2 mM EDTA. Transfer this into an inverted 5 ml sterile syringe using a 1 ml Pipetman.
4. Pass this solution through the cartridge and collect 1 ml fractions in microcentrifuge tubes (label the tubes '1' and '2').
5. Using a new syringe wash the column through with 5 ml DEPC water, collecting 1 ml fractions (label the tubes '3' through to '7').
6. Using a fresh 1 ml sterile syringe, elute the oligonucleotide by passing 1 ml of methanol/water (60:40) through the cartridge and collect the effluent in tube '8'. Repeat two more times to generate effluent in tubes '9' and '10'.
7. Measure the absorbance at OD260 nm of each of the ten tubes.[c] Take 50 μl of solution from each tube and dilute into 950 μl of methanol/water (60:40). Use methanol/water as a blank. There should be no absorbance in tubes 1–7. Most of the oligonucleotide should have eluted into tube '8', with some in tube '9' and very little in tube '10'. Discard all tubes except tube '8'. This tube becomes the 'methanol/water' stock.
8. If the oligonucleotide is a 45-mer, remove 0.5 μg of oligonucleotide in the corresponding volume of methanol/water (e.g. 200 μl) and dry this solution down under vacuum. Resuspend the oligonucleotide to 5 ng/μl

by adding 100 μl DEPC water. Store both the aqueous and methanol/water solutions at −20°C. The aqueous solution can be freeze-thawed whenever probe is required for labelling. Thaw out the methanol/water stock each time the aqueous supply becomes exhausted, and dry down more oligonucleotide.

[a] Suitable cartridges are 'SEP-PAK C_{18} cartridges for rapid sample preparation' (Waters Associates). Put the syringes into the long end of the column.
[b] It is sensible to use only a quarter to half of the total amount of the oligonucleotide obtained from the synthesizer, in case of accidents!
[c] For 1 ml of an oligonucleotide solution, the absorbance at 260 nm multiplied by 20 gives an answer in terms of μg/ml (ng/μl). Use silica cuvettes.

3. Labelling of probes

Probes suitable for *in situ* hybridization require high specific activities, allowing them to be used at very low concentrations (which will minimize non-specific binding) and allowing for short exposure times. Tailed oligonucleotides meet these requirements. The isotope used for tailing is either ^{32}P or ^{35}S. ^{32}P-labelled probes are used for Northern blot analysis and fast regional localization of transcripts in tissue sections on X-ray film. ^{35}S-labelled probes can be used for higher resolution X-ray film analysis and for single cell resolution with photographic emulsion.

3.1 The terminal transferase reaction

There are two common ways of labelling synthetic oligonucleotide probes. These are:

(a) *Kinasing*; in which [γ^{32}P]ATP is added to the 5'-end of the molecule using a kinase enzyme. This results in the addition of only one ^{32}P molecule to the oligonucleotide. Such probes are not usually very useful for *in situ* hybridization since the probe specific activity is too low.

(b) *Tailing with terminal deoxynucleotidyl transferase* (13, 14). This results in the addition of multiple residues of 5'-[α-^{32}P]dAMP or 5'-[α-^{35}S]dAMP to the 3'-end of the molecule. The number of residues added to the 3'-end depends on the molar ratio of oligonucleotide to isotope. The tail length, and hence the probe specific activity, can be controlled by simply changing this ratio. We have found that for *in situ* hybridization, a 1:30 molar ratio of probe to isotopic base in the labelling reaction is optimal (see Section 5). Increasing the tail length using for example a 1:90 ratio results in heavy non-specific binding. Adenosine residues are the preferred base to tail with, since in theory they will give less nonspecific binding than say G or C residues because they can form fewer hydrogen bonds. Tailing with T residues may result in hybridization to the poly [A] tail of mRNAs, and is therefore to be avoided.

Protocol 2. Labelling 45-mer oligonucleotide probes with terminal transferase

1. Make up a 12.5 µl reaction volume by adding the reagents in the order listed in this protocol. The 10 × tailing buffer[a] is pre-aliquoted in 1.25 µl aliquots in sterile 1.5 ml microcentrifuge tubes and stored at −20°C. For each raction, an aliquot of the buffer is thawed out and the other reagents are added directly to it. It is important not to introduce air bubbles or vortex/spin the tube, since the enzyme is inhibited by air bubbles.

2. For the preparation of ^{32}P-labelled probes, add to the 1.25 µl of 10 × buffer, 1 µl of oligonucleotide (5 ng/µl), 4.25 µl of DEPC treated water, 5 µl of 5'-[α-^{32}P]dATP (6000 Ci/mmol, 20 µCi/µl),[b] and 1 µl of terminal deoxynucleotidyl transferase (20 U/µl).[c] Mix gently by pipetting up and down. Incubate at 30–35°C for 1 h.

3. For the preparation of ^{35}S-labelled probes, add to the 1.25 µl of 10 × buffer, 8.25 µl of DEPC treated water, 1 µl of 5'-[α-^{35}S]dATP (1200 Ci/mM, 12.5 µCi/µl),[b] and 1 µl of terminal deoxynucleotidyl transferase (20 U/µl).[c] Mix gently by pipetting up and down. Incubate at 30–35°C for 1 hour.

4. Stop the reactions by adding 40 µl of DEPC treated water.

5. Unincorporated nucleotides are separated from labelled probe by a spin column procedure using Sephadex G-25.[d] Pipette the 50 µl of probe solution to the top of the pre-spun column, and spin at 2000 r.p.m. for 1 min, collecting the 50 µl of probe in a sterile microcentrifuge tube.

6. For ^{35}S-labelled probes, add 2 µl of 1 M DTT[e] to the eluate to prevent cross-linking of sulphur residues. Probes are now ready for use.[f]

7. Remove 2 µl of probe solution and transfer into scintillation vials for counting. For ^{32}P-labelled probes, sample can be counted directly without scintillation liquid (Cerenkov counting), and should be in the range 0.5–2.0 × 10^6 c.p.m./µl. For ^{35}S-labelled probes, add scintillant, and vortex the sample. The counts should be in the range 0.05–0.2 × 10^6 d.p.m./µl.

8. Dilute the probes in hybridization buffers.[g] Use 1 µl of probe (from the 50 µl eluate of the spin column) per 100 µl of hybridization buffer. The probe is used at a concentration of 1 pg/µl. For ^{35}S probes, add 20 µl of 1 M DTT per 100 µl of hybridization buffer. Vortex well. Do not boil the probe/hybridization buffer prior to use.

9. Proceed to *Protocol 8*.

[a] 10 × Tailing buffer is 1 M K Cacodylate, 10 mM $CoCl_2$, 20 µM DTT. K Cacodylate is prepared as described in ref. 16.
[b] 5'-[α-^{32}P]dATP and 5'-[α-^{35}S]dATP are obtained from NEN, Dupont.
[c] Terminal Deoxynucleotidyl Transferase (FPLC pure), Pharmacia.

^{d} 1 ml Sephadex G-25 spin columns are prepared exactly as described in ref. 19, except that the columns are pre-spun at 2000 r.p.m. (1 min) and spun with probe at 2000 r.p.m. (1 min).
^{e} Dissolve 3.09 g DTT in 20 ml DEPC treated water. Store in 1-ml aliquots at $-20°C$.
^{f} It is probably best to use ^{32}P-probes the same day, since radiolysis and probe break-up will be significant. ^{35}S-probes can be stored for longer periods at $-20°C$, at least for a week and probably longer.
^{g} For hybridization buffer, see *Protocol 5*.

The procedure for tailing 45-mer oligonucleotides is given in *Protocol 2*. When using 5'-[α-^{35}S]dATP with the tailing buffer, it is important to use a commercial source of isotope which has a low concentration of dithiothreitol (DTT), i.e. 1 mM. Isotope preparations which have high DTT concentrations to stabilize them (e.g. 10 mM) do not work well with the buffer. Excess DTT can form an insoluble pink complex with the cacodylate buffer which inhibits the reaction. When ^{32}P is used, Perspex shielding should be used, since the amount of radiation is considerable.

The specific activity of the resulting probe after the spin column is calculated by assuming the tail contributes radioactivity, but no mass (14). Using the conditions in *Protocol 2* for a 45-mer, 5 ng of labelled oligonucleotide will be present in 50 µl of eluate from the column. ^{32}P-Labelled probes normally have between 0.5×10^6–1.5×10^6 c.p.m./µl (as measured by Cerenkov counting), while for ^{35}S-labelled probes the value is usually between 0.1×10^6–0.2×10^6 d.p.m./µl (as measured by scintillation counting). These values translate to an average specific activity of 10^{10} c.p.m./µg for ^{32}P-labelled probes or 10^9 d.p.m. µg for ^{35}S-labelled probes. It has been demonstrated that the length of the tail does not affect the hybridization specificity (14), although longer tails may decrease hybridization efficiency (14).

3.2 Analysis of labelled probes on acrylamide gels

To check the actual length of the tail added to the oligonucleotide, it is advisable to run the probe on a 10% acrylamide/8 M urea gel [preparation of the gels is described elsewhere (15)]. *Figure 2* shows acrylamide gel analysis of oligonucleotides tailed with 5'-[α-^{32}P]dAMP and 5'-[α-^{35}S]dAMP. The 42-mer was incubated with a 30:1 ratio of labelled nucleotide to oligonucleotide, whereas the 48-mer was incubated with a 90:1 ratio of nucleotide to oligonucleotide. For each labelled probe, there is a population of molecules with different tail lengths, which appears to fit a normal distribution. Each band on the ladder represents one nucleotide difference (*Figure 2*).

4. In situ hybridization

All solutions used prior to and during the hybridization step must be sterile and free of contaminating ribonculeases. Solutions are sterilized by treating with diethylprocarbonate (DEPC). This compound is an alkyalting agent and

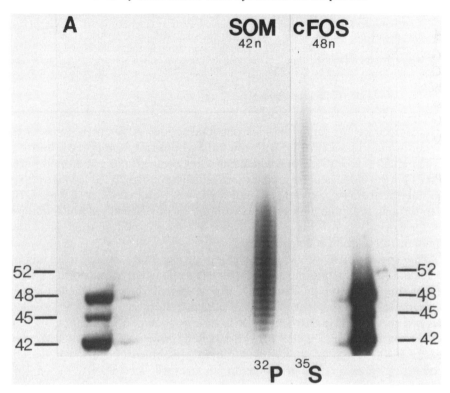

Figure 2. Analysing labelled probes on 10% acrylamide/8 M urea gels. The SOM oligonucleotide, a 42-mer, was incubated with a 30:1 molar ratio of 5'-[α-^{32}P]dATP. The cFOS oligonucleotide, a 48-mer, was incubated with a 90:1 molar ratio of 5'-[α-^{35}S]dATP. Oligonucleotides of known size (52, 48, 45, and 42 nucleotides long), were kinased and used as markers. See Section 3.2. (W. Wisden, B. J. Morris, and S. P. Hunt, unpublished data.)

consequently inactivates any protein present. Add 1.0 ml of diethylpyrocarbonate/litre of solution, shake vigorously and allow to stand for several hours. Do this in a fume hood since DEPC is a potential carcinogen. Then autoclave the solution which should destroy the DEPC (it breaks down into carbon dioxide and ethanol). Do not add DEPC to solutions containing tris buffers, since it reacts with primary amines. All glassware used for fixing and pre-treating sections should be baked overnight (this includes microscope slides, microscope slide racks, and siliconized coverslips). Gloves should be worn at all stages prior to hybridization to avoid contaminating samples with human ribonuclease! After the hybridization step, the DNA/RNA hybrids are resistant to ribonuclease (RNase), so that non-sterile conditions may be used.

4.1 Materials and reagents

Glassware. Glassware should be capable fo withstanding baking at 180°C. Continental glass troughs (100 × 90 × 70 mm deep) which are capable of holding 250 ml of solution. Glass racks (20-slide capacity). Glassware supplied by Arnold R. Horwell Ltd, London, UK. Siliconized glass coverslips (*Protocol 3*), Poly-L-lysine coated slides (*Protocol 6*). Hybridization buffer (*Protocol 4* and *5*).

Protocol 3. Cleaning glass coverslips for *in situ* hybridization

1. Take a packet of coverslips[a] and individually separate them. Place them into boiling 0.1 M HCl for 20 min. Do this in a fume hood. Allow to cool, and then pour off the acid.
2. Rinse the coverslips in deionized water and allow them to dry individually standing upright in a rack.
3. Siliconize the coverslips (in fumehood) by dipping them individually in dimethyldichlorosilane solution.[b] Allow to air-dry in rack.
4. Transfer the coverslips into a baking resistant glass Petri dish, and wash them thoroughly with several rinses of deionized water.
5. Wrap the Petri dish in aluminium foil and bake at 180°C overnight.

[a] Microscope glass coverslips (Chance Propper Ltd, No. 1, 22 × 50mm).
[b] Dimethyldichlorosilane solution (2% in 1,1,1 trichloroethane solution BDH).

Protocol 4. Preparation of acid/alkali cleaved salmon sperm DNA

1. Transfer 1 g salmon sperm DNA[a] to 50 ml sterile polypropylene tube. Add 15 ml of DEPC water and allow to soak for 15 min to 2 h.
2. Add 2.5 ml 2 M HCl, keep the DNA at room temperature. The DNA forms a white precipitate. Shake well until the precipitate sticks together. Gather into a ball with a pipette tip over 2 to 3 min.
3. Add 5.0 ml of 2.0 M NaOH. Shake to resuspend the DNA which should dissolve. Place the tube at 50°C for 15 min to increase dissolution.
4. Dilute the mixture to 175 ml with DEPC water, making sure there are no particles.
5. Add 20 ml of 1 M Tris–HCl, pH 7.4.
6. Titrate with 2 M HCl until the pH of the DNA solution reaches 7.5–7.0.
7. Filter the solution through a sterile millipore filter to remove any particles. Measure the OD260 nm absorbance of the solution. Pipette

In situ hybridization with synthetic DNA probes

20 μl of DNA solution into 980 μl of water. The absorbance multiplied by 50 gives the concentration of DNA in μg/ml.

8. Freeze the solution at −20°C, and freeze thaw stock as required. Do not boil before use.

[a] Deoxyribonucleic acid sodium salt type III, from salmon sperm testis.

Protocol 5. Oligonucleotide hybridization buffer

The hybridization buffer is 50% formamide, 4 × SSC, 10% dextran sulphate, 5 × Denhardts, 200 μg/ml acid–alkali cleaved salmon sperm DNA, 100 μg/ml long chain polyadenylic acid, 120 μg/ml heparin, 25 mM sodium phosphate pH 7.0, 1 mM sodium pyrophosphate.

In a sterile 50 ml polypropylene tube, add the following;
  25 ml of 100% deionized formamide[a]
  10 ml of 20 × SCC[b]
  2.5 ml of 0.5 M sodium phosphate pH 7.0[c]
  0.5 ml of 0.1 M sodium pyrophosphate
  5 ml of 50 × Denhardts solution[d]
  2.5 ml of 4 mg/ml acid–alkali hydrolysed salmon sperm DNA[e]
  1 ml of 5 mg/ml polyadenylic acid[f]
  50 μl of 120 mg/ml heparin[g]
  5 g of dextran sulphate[h]
  Adjust to 50 ml with DEPC treated water

The dextran sulphate takes several hours to completely dissolve with occasional vortexing.

Store the hybridization buffer at 4°C wrapped up in foil. It keeps for several months. Do not boil the hybridization buffer prior to use.

[a] Add 50 ml formamide (Fluka) per 5 g of BDH 'Amberlite' monobed resin MB-1. Stir at room temperature for half an hour. Filter through Whatman 3 mm paper and store at −20°C in 25 ml aliquots.

[b] 20 × SSC is 3 M NaCl, 0.3 M Na citrate, pH 7.0 with HCl. Filtered, DEPC treated and autoclaved.

[c] 0.5 M sodium phosphate pH 7.0 is prepared by mixing 0.5 M Na_2HPO_4 and 0.5 M NaH_2PO_4 until the pH reaches 7.0. The solution is then filtered, DEPC treated and autoclaved.

[d] 50 × Denhardt's is 5 g polyvinylpyrrolidine (PVP, 5 g bovine serum albumin (BSA), 5 g Ficoll 400/500 ml of DEPC treated water. Store in aliquots at −20°C.

[e] Prepared as described in *Protocol 4*.

[f] Dissolve 100 mg of polyadenylic acid [5'], potassium salt (Sigma No.P-9403) in 20 ml of DEPC treated water, to give 5 mg/ml stock. Store in 1 ml aliquots at −20°C.

[g] Heparin, sodium-freeze dried (made by BDH) dissolved up to 120 mg/ml in DEPC treated water and stored in 50 μl aliquots at −20°C.

[h] Dextran sulphate, sodium salt (Pharmacia).

4.2 Preparation and fixation of tissue

Tissue (e.g. rat brain) is dissected out from the animal fresh (i.e. not perfusion fixed), preferably within several hours of death. Most RNA appears to be stable in intact brain at room temperature for at least 12 h after death (17), since most cellular RNA is probably compartmentalized away from ribonucleases. This is useful to know when collecting post-mortem tissue. Degradation only becomes a problem when the tissue has been frozen and then allowed to thaw, since the cells then lyse allowing cellular contents to mix. Once the tissue has been removed and dissected, it is placed on a strip of aluminium foil resting on dry ice. The tissue freezes completely within minutes. The tissue is then mounted onto a cutting block with a cryoglue, and transferred to the cryostat chamber at $-20°C$ to equilibrate prior to cutting. Alternatively, once the specimen has been frozen on dry ice, it is wrapped in parafilm (to prevent freeze drying), and stored at $-70°C$. Tissue can be kept in this way for very long periods of time (months to years?). The actual mechanics of sectioning tissue are not covered in this chapter.

The method is described in *Protocol 6*. The tissue is sectioned at between 12–15 μm, and is then thaw-mounted onto sterile poly-L-lysine coated slides. Wear gloves throughout this procedure. The preparation of slides is given in *Protocol 7*. Sections are then allowed to air-dry at room temperature for anywhere between half an hour to several hours. The sections are usually completely dry within minutes. The slides are then placed in a carrier and immersed in ice-cold buffered paraformaldehyde (4% w/v) for 5 min. This is a very light fixation step but is sufficient to retain RNA in the section. The sections are then briefly rinsed in phosphate buffered saline (PBS) before being dehydrated in 70% ethanol. Most protocols usually air-dry the sections at this point before storing them dessicated at $-70°C$. However, we have found it more convenient to store sections in 95% ethanol at 4°C (B. Morris, unpublished observations). Stored in this way, the signal does not appear to diminish even after six months, and one does not have to wait several hours while frozen sections are allowed to warm up to room temperature. Sections stored in ethanol can be removed and used immediately. (In addition, ethanol storage makes it much easier to reorganize and see exactly what samples are available for use). Ethanol storage may also remove a substantial amount of lipid from the sections (white matter is sometimes a source of probe non-specific binding). **It is important to remember not to store 95% ethanol in the 4°C refrigerator since ethanol is highly flammable and could explode if ignited by a spark from a thermostat. Instead, samples should be stored in a cold room.**

Protocol 6. Preparation and fixation of sections

1. Cut 12–15 μm sections on a cryostat at −20°C. Thaw mount the sections onto poly-L-lysine coated slides.[a] Allow sections to dry at room temperature for half an hour to several hours.
2. Prepare a 4% solution of depolymerized paraformaldehyde (PFA) as follows. Transfer 40 g paraformaldehyde[b] into 500 ml of DEPC treated water. Heat the milky white suspension with continuous stirring until it reaches 60–65°C. Do not heat above this temperature. Do the whole procedure in a fume hood. Add 1.0 M NaOH dropwise until the suspension clears. Add 500 ml of 2 × PBS.[c] Mix well and chill the solution in an ice/water bath. Check that the pH is roughly 7.0. Use the same day.
3. Transfer a rack of dry sections into the ice-cold 4% PFA. Leave for 5 minutes.[d]
4. Transfer sections into 1 × PBS[c] for a minute.
5. Transfer sections into 70% ethanol[e] for several minutes.
6. Transfer the sections into the storage box containing 95% ethanol. Store at 4°C in cold room until required.

[a] See *Protocol 7* for the preparation of poly-L-lysine coated slides.
[b] Paraformaldehyde (powder) is general purpose reagent grade (BDH).
[c] 10 × PBS is 1.3 M NaCl, 70 mM Na_2HPO_4, 30 mM NaH_2PO_4. Filter, DEPC treat and autoclave. 2 × PBS and 1 × PBS are prepared by diluting the 10 × stock with DEPC treated water.
[d] We use 20 slides/glass rack and 250 ml of solution in continental staining troughs. See Section 4.1.
[e] Diluted from 100% ethanol with DEPC treated water.

Protocol 7. Preparation of poly-L-lysine coated slides

1. Dissolve 25 mg of poly-L-lysine hydrobromide (MW 35 000)[a] in 5 ml of DEPC treated water. Store as 1 ml aliquots at −20°C.
2. Thaw out 1 ml of the 5 mg/ml aliquots of poly-L-lysine, and dilute it to 50 ml with DEPC treated water, to give a 0.01% solution. Transfer this solution to a sterile 50 ml disposable plastic Petri dish.
3. Dip the baked slides[b] individually in the poly-L-lysine solution (immerse each slide completely), and allow slides to air-dry standing upright in a rack. Use slides preferably on the same day, although they can be stored at 4°C for a limited period.

[a] Poly-L-lysine hydrobromide, MW 350 000 (Sigma, P-1524).
[b] A packet of microscope slides is wrapped in aluminium foil and baked for 4 h to overnight at 180°C. Slides are allowed to cool to room temperature before use.

We store sections in plastic air-tight boxes which hold approximately two litres of ethanol with six racks of sections (20 slides/rack). The lids are sealed with a layer of insulation tape as a further precaution to prevent evaporation (if the sections dry out for long periods, the signal will be substantially reduced).

4.3 Hybridization of probes to tissue

Just prior to hybridization, sections are removed from the 95% ethanol and allow to air-dry (*Protocol 8*). We and others (8), have found that no pretreatment of sections is necessary (see Section 5). However, if one has problems with probes binding to white matter, then the procedure given in *Protocol 9* may be used. In this case, sections are first transferred straight from the ethanol into PBS. They are then acetylated (this step coats the section with negative charge which may repel negatively charged nucleic acid probes), followed by a delipidation step in chloroform.

Protocol 8. Application and hybridization of probes to sections

1. Remove the sections from 95% ethanol storage and allow to air-dry for 15 min to half an hour.
2. Dissolve the probe in hybridization buffer.[a] 60 µl of hybridization buffer per 20 × 55 mm coverslip is required.
3. Apply 60 µl of probe/hybridization buffer to each slide. If several sections are on each slide, make sure that there is some hybridization buffer on all the sections.
4. Gently lower a siliconized coverslip over the drop of hybridization buffer. The liquid should spread smoothly under the coverslip. Remove any air bubbles by very gentle pressing with forceps.[b]
5. Seal the coverslips with melted paraffin wax.[c] Use a pre-warmed pipette to trail melted wax around the edge of the coverslip. As soon as it touches the slide it sets. The wax is kept liquid on a hotplate.
6. Saturate a small piece of tissue/filter paper with 50% formamide/4 × SSC and place this in the Petri dish as well. Place the lid on the dish and incubate overnight at 42°C.

[a] See *Protocol 2* (step 8).
[b] Slides are laid horizontally in a disposable transparent plastic Petri dish. If this is placed on a black surface it becomes easy to see any air bubbles. Coverslips are prepared as described in *Protocol 3*.
[c] Paraffin wax (BDH, congealing point about 60°C, pastilated).

Protocol 9. Optional pre-treatment of sections[a]

For all the steps listed below, used baked glassware and DEPC treated water to make up all solutions.

1. Transfer sections straight from 95% ethanol into 1 × PBS.
2. Transfer into 0.25% acetic anhydride in 0.1 M triethanolamine-HCl pH 8.0/0.9% NaCl for 10 min at room temperature.
3. Transfer into 70% ethanol for 1 min.
4. Transfer into 95% ethanol for 2 min.
5. Transfer into 100% ethanol for 1 min.
6. Transfer into chloroform (100%) for 5 min.
7. Transfer back through 100% (1 min) and 95% ethanol (1 min).
8. Air-dry the sections and proceed to *Protocol 8*, step 1.

[a] This protocol is used by Young *et al.* (7).
[b] To prepare the acetic anhydride solution, add 3.3 ml of triethanolamine (it comes as a viscous liquid), 1.25 g of NaCl and 1.0 ml of conc. HCl/250 ml DEPC water. Mix well. Just prior to use add 625 μl of acetic anhydride. Stir well with a sterile pipette tip.

Once the sections are dry (10 min), the probe dissolved in hybridization buffer is applied directly to them. They are then covered with siliconized glass coverslips and incubated at 42°C overnight. The coverslips are sealed with melted paraffin was to prevent drying out. Alternatively, parafilm can also be used instead of (in some ways more cumbersome) glass coverslips (*Protocol 8*). Parafilm requires no wax sealing step.

The next day (see *Protocol 10*), coverslips are removed individually from the sections by immersing each slide in room temperature 1 × SSC (containing 20 mM DTT if ^{35}S-labelled probes are being used). Coverslips are gently teased off using blunt-ended forceps. The rack of sections is then transferred to a pre-warmed solution of 1 × SSC at 55°C. This is best performed by placing the continental troughs in a gently agitating water bath. Wash times can be a short as 20 min. No change of SSC solution is necessary. After the high temperature wash, slides are taken rapidly through room temperature 1 × SSC, room temperature 0.1 × SSC, 70% ethanol and finally 95% ethanol. They were then allowed to air-dry. We now routinely omit DTT from wash solutions as this was found to be unnecessary.

Protocol 10. Washing sections after the hybridization

All procedures can be non-sterile at this point.

1. Transfer the slides (still with the coverslips on) into 1 × SSC at room temperature. If ^{35}S-probes are being used, add 1 ml of 1 M DTT per 250 ml of wash solution. Gently dislodge the coverslips with blunt-ended

forceps. The coverslips should float to the surface still attached to the wax. Transfer the slides and remove the coverslips one at a time.

2. Transfer the rack of slides into 205 ml pre-warmed 1 × SSC at 55°C. Add 1 ml of 1 M DTT for ^{35}S-probes. It is convenient to have the troughs in a shaking water bath. Leave the sections washing for half an hour.

3. Transfer the rack through a very brief (couple of seconds) series of room temperature rinses in 1 × SSC, 0.1 × SSC, 70% ethanol, 95% ethanol. Allow sections to air-dry.

4. Expose sections to X-ray film or dip in emulsion.

4.4 Autoradiography of sections

The type of autoradiographic procedure used depends on the isotope (i.e. ^{32}P or ^{35}S). For an explanation of the principles of autoradiography, consult Bonner (18). If slides have been hybridized with ^{32}P-labelled probes, then a scan of the section with the Geiger counter usually reveals a significant number of counts above background with respect to control sections. They can then be exposed to X-ray film. We typically use two types of film. They are (a) XAR-5 (Kodak) which is a fast very sensitive low resolution film, and (b) RX-100 (Fuji) which is slower and less sensitive, but gives higher resolution images. For an initial examination of the sections, they can be exposed at −70°C with scintillation screens using XAR-5 film. This gives a quick low resolution signal. It is important to allow the cassette to warm up to room temperature before development, otherwise water will condense on to the slides. Specimens can be exposed for ten times longer without screens at room temperature using Fuji Rx-100. It is important to use tight fitting cassettes that press the slides firmly against the X-ray film, thus generating on even pressure. This will avoid 'out of focus' images. We have found it best to use Kodak X-Omatic cassettes fitted with Super-rapid screens. In general, ^{32}P-labelled probes do not work efficiently with emulsion, although if target message levels are particularly high, a signal may be detected (see *Figure 3B*).

When ^{35}S-labelled probes are used, it is usually very difficult to detect counts on the sections using a Geiger counter. Specimens should be exposed to XAR-5 at room temperature (all β electrons emitted by ^{35}S are blocked by film, so that nothing will be gained by exposing the sections at −70°C, which is the temperature that scintillation screens work most efficiently). Sections can then be dipped in photographic emulsion. Details of this procedure are given elsewhere (20, 21).

5. Controls for *in situ* hybridization

The patterns of probe binding to brain sections can be subject to some well documented artefacts. For example, areas of high cell density can tend to

In situ hybridization with synthetic DNA probes

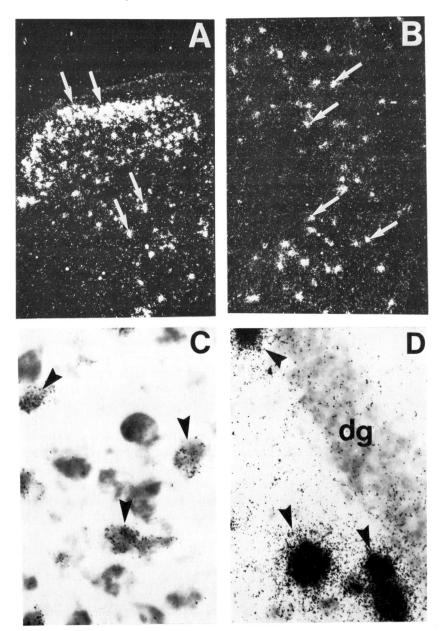

Figure 3. Illustrations of single-cell resolution obtained by coating sections with photographic emulsion. **A**, Localization of preprotachykinin mRNA in the dorsal horn of the rat spiral cord using a ^{35}S-labelled antisense oligonucleotide probe. Dark-field photomicrograph. Labelled cells are marked by clusters of silver grains. Exposure time, 4 weeks. **B**, Localization of the GABA$_A$ receptor α_1 subunit mRNA in the bovine substantia

bind probe indiscriminately. Such areas include the dentate gyrus of the hippocampus, the granule cell layer of the cerebellum, the olfactory bulb granule cell layers, the habenula, and white matter. However, using the conditions described here, non-specific binding in these areas does not usually occur. Nevertheless, patterns of labelling must be carefully interpreted with respect to controls. The types of controls are outlined below, and some of them are illustrated in *Figure 4*.

(a) It may be useful to use several different probes built to different regions of the mRNA. These should give identical patterns of hybridization (assuming there is no alternative splicing).

(b) Sense probes of the same length and GC content should not detect any signal.

(c) Incubating the sections with a large excess (say 20-fold) of unlabelled probe combined with the normal concentration of labelled probe should only compete out specific binding.

(d) The probe should be tested on a Northern blot to make sure it hybridizes to the correct transcript. Hybridization and wash conditions should be identical to those used for *in situ* hybridization, except that dextran sulphate is omitted from the hybridization buffer (since it tends to cause excessive filter background), and the probe is used at a much lower concentration (10^6 c.p.m./ml of ^{32}P-tailed probe).

(e) To test that the pattern is the result of probe hybridizing to RNA, some sections should be pre-treated with Ribonuclease A (RNAseA) prior to hybridization. To do this, sections are removed from 95% ethanol and transferred into 2 × SSC containing 20 µg/ml of RNaseA. Sections are incubated at 37°C for 30 min. They are then dehydrated and hybridized as normal. It is important to keep a separate pipette aside solely for the purposes of pipetting RNAse solutions. All glassware used for RNAase pre-treatments should be kept separate to avoid contamination.

In initial experiments, we found that trying to increase the specific activity of ^{35}S-labelled probes by using a very long tail length (a 90:1 molar ratio of isotopic base to oligonucleotide was used in the tailing reaction), resulted in

---

nigra using a ^{32}P-labelled antisense oligonucleotide. Dark-field photomicrograph. Exposure time, 3 weeks. **C**, Localization of the avian nicotinic receptor α_4 subunit mRNA in the chick lateral spiriform nucleus using a ^{35}S-labelled oligonucleotide. Bright-field photomicrograph. Note that not all of the cells are labelled. Exposure time, 6 weeks. **D**, Localization of the mRNA encoding preprosomatostatin in the rat hippocampal hilar region. Note that the dentate gyrus (**dg**) is unlabelled. Exposure time, 4 weeks. Arrows and arrowheads indicate examples of labelled cells. The emulsion was Ilford K5, exposed at 4°C, developed with Kodak D19 (2 mins at 17°C) and fixed with 30% sodium thiosulphate (2 min). Sections were counterstained with thionin. (All from W. Wisden, B. J. Morris, and S. P. Hunt, unpublished data.)

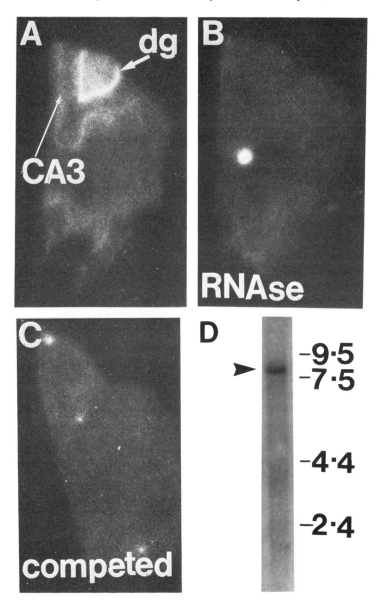

Figure 4. Demonstration of probe specificity. **A**, Film autoradiograph showing regional localization of the GABA$_A$ receptor α_2 subunit mRNA in the bovine hippocampal complex using a ^{32}P-labelled antisense 45-mer (**dg**, dentate gyrus). **B**, Serial section that was pre-incubated with RNseA (see Section 5) prior to probe application. **C**, Serial section in which the hybridization step took place with the same concentration of labelled probe that was used in **A**, but in addition the hybridization buffer also included a 16-fold excess of unlabelled antisense α_2 oligonucleotide. Exposure time for **A**, **B**, and **C** was overnight (with screens at −70°C). **D**, Northern blot demonstrating that the α_2 oligonucleotide

very heavy non-specific binding. A 30:1 molar ratio was found to be optimal (W. Wisden, B. J. Morris, and S. P. Hunt, unpublished observations). In addition, we found that whilst our original experiments were conducted without dextran sulphate in the hybridization buffer, a significant increase in the signal-to-noise ratio was observed when dextran sulphate was added to 10% w/v in the hybridization buffer. This may be due to the fact that without dextran sulphate present, the hybridization reaction may not go to completion. In addition, the pre-treatments of the sections given in *Protocol 9* were found to make no difference to the signal-to-noise ratio, enabling this step to be routinely omitted. These points are demonstrated in *Figure 5*.

6. Prospects

There are now several reports of using non-radioactively labelled oligonucleotide probes for *in situ* and membrane hybridizations (22, 23, 24). The potential advantages of these systems is that they offer much higher resolution results which can be obtained quickly, and without the hazards of working with radiation. However, these methods currently appear to be much less sensitive than radiolabelled probes, although this is clearly a growth area and many improvements to the technique can be expected.

Acknowledgements

We thank Dr M. Goedert (L.M.B., Cambridge) for the methods in *Protocol 1* and for much useful advice, Dr O. Sundin for the protocol to prepare acid/alkali hydrolysed salmon sperm DNA, for D. Nunez (AFRC Institute, Babraham, Cambridge) for advice on preparation of the tailing buffer, Drs S. Davies, M. Spillantini, M. Darlison and Prof. E. Barnard for discussions, Dr A. Gundlach for critically reading the manuscript, and Dr J. Davies (British Bio-technology, Oxford, UK) for supplying oligonucleotides. B. J. M. was supported by the Mental Health Foundation. W. W. held an MRC Research Studentship.

References

1. Goedert, M. and Hunt, S. P. (1987). *Neuroscience*, **22**, 983.
2. Garner, G. C., Tucker, R. P., and Matus, A. (1988). *Nature*, **336**, 674.
3. Morris, B. J., Herz, A., and Hollt, V. (1989). *Journal of Molecular Neuroscience*, **1**, 9.

---

detects an 8.0 kb transcript (arrowhead) in poly A^+ RNA (20 µg) isolated from bovine hippocampus. Size markers are indicated in kilobases. (W. Wisden, B. J. Morris, and S. P. Hunt, unpublished data.)

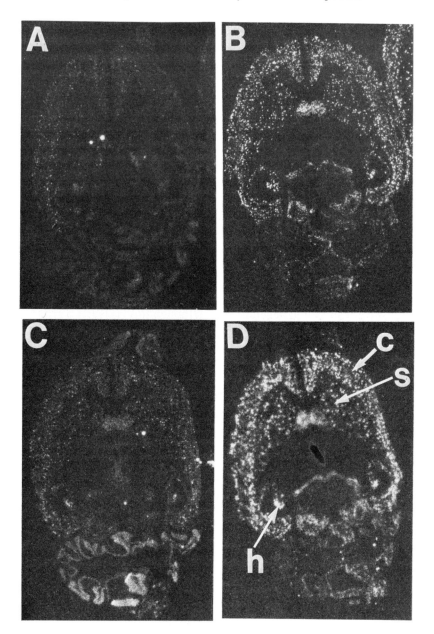

Figure 5. Comparison of section pre-treatments and hybridization buffer (with and without dextran sulphate), using a ^{35}S-labelled oligonucleotide probe recognizing rat preprosomatostatin mRNA. **A**, pre-treated section (*Protocol 10*), no dextran sulphate. **B**, pre-treated section (*Protocol 10*), 10% dextran sulphate in hybridization buffer. **C**, Non-pre-treated section, no dextran sulphate. **D**, non-pre-treated section, 10% dextran sulphate in hybridization buffer. Film autoradiographs were obtained by exposing the

4. Morris, B. J., Feasey, K. J., ten Bruggencate, G., Herz, A., and Höllt, V. (1988). *Proceedings of the National Academy of Sciences of the USA*, **85**, 3226.
5. Wilkinson, D. G., Bailes, J. A., and McMahon, A. P. (1987). *Cell*, **50**, 79.
6. Hunt, S. P., Wisden, W., Morris, B. J., Davies, S. W., Spillantini, M. G., and Goedert, M. (1989). In *Neuropeptides: A Methodology* (ed. G. Fink and A. J. Harmer), p. 55. Wiley, IBRO.
7. Young, W. S., Mezey, E., and Siegel, R. E. (1986). *Neuroscience Letters*, **70**, 198.
8. Lewis, M. E., Krause, R. G., and Roberts-Lewis, J. M. (1988). *Synapse*, **2**, 308.
9. Wisden, W., Morris, B. J., Darlison, M. G., Hunt, S. P., and Barnard, E. A. (1988). *Neuron*, **1**, 937.
10. Morris, B. J., Wisden, W., Dunnett, S. B., and Sirinathsinghji, D. J. S. (1989). *Neuroscience Letters*, **103**, 121.
11. Lathe, R. (1985). *Journal of Molecular Biology*, **183**, 1.
12. Albretsen, C., Haukanes, B. I., Aasland, R., and Klepper, K. (1988) *Analytical Biochemistry*, **170**, 193.
13. Deng, G. and Wu, R. (1981). *Nucleic Acids Research*, **9**, 3719.
14. Collins, M. L. and Hunsaker, W. R. (1985). *Analytical Biochemistry*, **151**, 211.
15. Ogden, R. C. and Adams, D. A. (1987). In *Methods in Enzymology*, Vol. 152 (ed. S. L. Berger and A. R. Kimmel), p. 61. Academic Press, London.
16. Eschenfeldt, W. H., Puskas, R. S., and Berger, S. L. (1987). In *Methods in Enzymology*, Vol. 152 (ed. S. L. Berger and A. R. Kimmel), p. 337 Academic Press, London.
17. Pittius, C. W., Kley, N., Loeffler, J. P., and Höllt, V. (1985). *EMBO Journal*, **4**, 1257.
18. Bonner, W. M. (1987) In *Methods in Enzymology*, Vol. 152 (ed. S. L. Berger and A. R. Kimmel), p. 55. Academic Press, London.
19. Maniatis, T., Fritsch, E. F., and Sambrook, J. (1982). *Molecular Cloning, A Laboratory Manual*, p. 466. Cold Spring Harbor Laboratory, Cold Spring Harbor, New York.
20. Uhl, G. R. (ed.) (1987). *In situ Hybridization in Brain*. Plenum Press, New York.
21. Valentino, K. L., Eberwine, J. H., and Barchas, J. D. (ed.) (1987). *In situ Hybridization, Applications to Neurobiology*. Oxford University Pres, Oxford.
22. Arai, H., Emson, P. C., Agrawal, S., Christodoulou, C., and Gait, M. J. (1988). *Molecular Brain Research*, **4**, 63.
23. Guitteny, A., Fouque, B., Mougin, C., Teoule, R., and Bloch, B. (1988). *Journal of Histochemistry and Cytochemistry*, **36**, 563.
24. Riley, L., Marshall, M., and Coleman, S. (1986). *DNA*, **5**, 333.

---

sections to XAR-5 at room temperature for 5 d (probe concentrations and exposure times were identical for all comparisons). The individual dots on the film correspond to labelled cells. **c**, cortex, **h**, hippocampus, **s**, striatum. (W. Wisden, B. J. Morris, S. P. Hunt, unpublished data).

Cellular neurobiology— a practical approach

CELLS *IN VITRO*

1. Culture of adult mammalian peripheral neurons
   Ronald M. Lindsay, Caroline J. Evison, and Janet Winter
2. Dissociated neurons from adult rat hippocampus
   J. E. Chad, I. Stanford, H. V. Wheal, R. Williamson, and G. Woodhall
3. Production of the organotypic slice cultures of neural tissue using the roller-tube technique
   Richard T. Robertson, Casey M. Annis, and Beat H. Gähwiler
4. Chemical synapses in cell culture
   P. G. Haydon and M. J. Zoran

WHOLE-CELL RECORDINGS AND ISOLATION OF IONIC CURRENTS

5. NG 108-15 neuroblastoma × glioma hybrid cell line as a model neuronal system
   R. J. Docherty, J. Robbins, and D. A. Brown
6. Isolation of potassium currents
   B. Lancaster
7. Perfusion of nerve cells and separation of sodium and calcium currents
   P. G. Kostyuk
8. Whole-cell voltage-clamp techniques applied to the study of synaptic function in hippocampal slices
   Daniel Madison

OPTICAL TECHNIQUES

9. Photolabile calcium buffers to selectively activate calcium-dependent processes
   Alison M. Gurney

10. Caged molecules activated by light
    Jeffrey W. Walker
11. Optical techniques and Ca^{2+} imaging
    R. B. Moreton

HEURISTIC MODELLING

12. Three-dimensional reconstructions and analysis of the cable properties of neurons
    D. Turner, H. V. Wheal, E. Stockley, and H. Cole
13. Modelling the non-linear conductances of excitable membranes
    Lyle Borg-Graham

Index

acetylcholine receptors (AChR)
  muscarinic 163
  nicotinic 60, 163
acrylamide gel electrophoresis
  labelled oligonucleotide probes 211, 212
  ratiolabelled phosphoproteins 137–8
activation
  Ca^{2+} channels 28, 29, 34–6, 39, 45
  glutamate-gated channels 63–4
ADP-ribosylation factor (ARF) 103
affinity chromatography
  anti-$GABA_A$ receptor antibodies 197–9
  calmodulin kinase II 145
AlF_4^- 101–2
amiloride 31
4-aminopyridine (4-AP) 33
AMPA (α-amino-3-hydroxy-5-methyl-4-isoxazoleproprionic acid) 52, 67
angiotensin II receptors 162
anti-$GABA_A$ receptor antibodies
  production 191–2
  purification using affinity chromatography 197–9
  screening methods 191–202
    ELISA and immunoblots 192–9
    soluble immunoprecipitation assay 199–202
  synthetic peptides as antigens 185–91
    choice of peptide sequence 185–8
    coupling of peptide to protein carrier 188–91
anti-G-protein antibodies 105–6, 108
L-AP4 (L-2-amino-4-phosphonobutanoic acid) 53
arsenazo III 66–7
aspartate 33, 58
ATP 54, 108
  ^{32}P-labelled ([γ-^{32}P]ATP) 136, 138, 140, 146–7, 148
autoradiography
  in situ hybridization tissue sections 219
  radiolabelled phosphoproteins 137, 138

baclofen 103, 104
bandwidth 8–9, 11–13
BAPTA 34, 46, 54
barium ions (Ba^{2+}) 29, 30, 32–3, 41
BayK 8644 30, 44, 46
benzodiazepines 183
big patch clamping technique, *Xenopus* oocytes 176, 177–8
bilayers, patch clamping 11–12, 15

bis-diazo-*o*-tolidine method for coupling of peptide and carrier protein 189–90
Boltzmann distribution
  Ca^{2+} channels 37, 39
  stretch-activated ion channels 83–4
bombesin receptors 162
botulinum toxin 96, 102, 103
brain protein tyrosine phosphatase 5 (PTPase 5) 153
brain slices, *see* tissue slices

cadmium ions (Cd^{2+})
  glutamate receptor interactions 59
  sensitivity of Ca^{2+} channels 29, 31, 39
caesium ions (Cs^+) 33, 54
caged guanine nucleotide analogues 108–9
calcineurin 136, 149–50, 152, 153
  assay 147–9
  purification 150
calcium (Ca^{2+}) channels 27–46
  distinguishing features of different types 28–31
    kinetic properties 28–30
    permeability 30
    pharmacology 30–1
  G protein-coupled 99, 106, 110
  L-type 28–31, 34–5, 38, 44, 45–6
  N-type 28–31, 38, 45–6
  rundown 34, 36
  single-channel current recordings 29, 32, 40–5
    bath solution 32, 42
    conductance 42, 45
    kinetic analysis 45
    pharmacology 42–5
    pipette solution 32, 41
    pulse protocols 42
  supplies of chemicals 46
  T-type 28–31, 38, 41, 46
  whole-cell current recordings 31–40
    activation plots 39
    extracellular solutions 32–3
    inactivation plots 39
    intracellular solutions 33–4
    measuring rates of decay 40
    permeability 36
    pharmacology 38–9
    pulse protocols 34–6
    strategy for current isolation 39
  Xenopus oocytes
    endogenous 178–9
    exogenous 162
calcium chelators (buffers) 34, 68

Index

calcium indicators 66–8
calcium ions (Ca^{2+})
  glutamate (NMDA) receptors and 49, 54, 56, 58–9, 66–8
  growth cone motility and 76
  intracellular messenger function 135
  in solutions for studying Ca^{2+} channels 32–3, 41
  volume regulation and 88
calmodulin 135–6, 144, 152
calmodulin kinase I 141
calmodulin kinase II (multifunctional calmodulin-stimulated protein kinase) 136, 141, 143–5
  assay 144
  primary structure of substrate 143
  purification 144–5
calmodulin kinase III 141
capacitance, membrane, measurement of changes 15–16
capacitor feedback technique 9–15
  noise, dynamic range, linearity and bandwidth 11–13
  pros and cons 14–15
  rationale 9–11
  resets 13–14
  speed of observed traditions 14
  transients 14
casein kinase I 141
casein kinase II 141, 143
cDNAs 161
  RNA synthesis from 162, 166–7
cell cultures, *see* glial cells, primary cultures; neurons, primary cultures
cell growth regulation 76
cerebellar granule cells, glutamate receptors 58–9, 60
cerebral cortical slices, agonist-induced InsP formation 118–19, 120–1, 123, 124
CGP 28392 30
chloride channels 69, 179
chloride ions (Cl⁻) 33, 41, 85, 88
chloroform/methanol extraction, inositol phosphates 125
cholera toxin 96, 102, 103–5
choline 33
choroid plexus, agonist-induced InsP formation 123, 124
CNQX (6-cyano-7-nitro-quinoxaline-2,3-dione) 53, 66
concentration clamp techniques, glutamate-gated channels 61–4
conductances, membrane 4
  Ca^{2+} channels 29, 30, 42, 45
  glutamate-gated 55–6, 58–60
  stretch-activated ion channels 81
correction (positive feedback) technique of series resistance compensation 16–18

coverslips, cleaning, for *in situ* hybridization 213
crayfish
  muscle 63
  stretch receptors 76, 81–4, 85, 89
current, membrane 4
current-to-voltage (resistor feedback) patch-clamp headstage 7–9
current-voltage relationship (I–V)
  Ca^{2+} channels 39
  glutamate-gated channels 55–6
  stretch-activated ion channels 84, 85
cyclic AMP 34, 135
cyclic AMP-dependent-kinase (PKA) 141, 143
cyclic GMP-dependent-kinase 141
cytochalasin B 81, 89
cytoskeleton, stretch-activated ion channel activation and 89

deactivation (tails), Ca^{2+} channels 29–30, 36
decay times
  Ca^{2+} channels 28–9, 40
  EPSCs 63–4
desensitization, non-NMDA channels 63–4
dextran sulphate, hybridization buffer 214, 221, 223, 224
diacylglycerol (DAG) 115, 130–
diethylprocarbonate (DEPC) 211–12
digital signal processors (DSPs) 9
dihydropyridine (DHP) receptors, skeletal muscle 162
dihydropyridines (DHP)
  sensitivity of Ca^{2+} channels 29, 30–1, 38, 42–4
  sources 46
diphtheria toxin 102
dithiothreitol (DTT) 211
DNA
  probes for *in situ* hybridization 205
  salmon sperm, preparation 213–14
DNQX (6,7-dinitro-quinoxaline-2,3-dione) 5
dorsal root ganglion (DRG) neurons, Ca^{2+} channels 28–31, 32, 110
Dowex anion-exchange columns, *see* ion exchange chromatography

EDTA 139, 163
EGTA 34, 54, 68
elastic area compressibility modulus (k_s) 77–80–1
electroporation 108
ELISA (enzyme-linked immunosorbent assay), screening of anti-GABA$_A$ receptor antibodies 192–5, 197
enzymes, G protein-linked 97, 99
Escherichia coli, stretch-activated ion channel 81

230

Index

ethanol, storage of *in situ* hybridization sections 215
1-ethyl-3-(3-dimethylaminopropyl)-carbodiimide (EDAC) method for coupling of peptide and carrier protein 190–191
excitatory amino acid (EAA) receptors, *see* glutamate receptors
excitatory post-synaptic current (EPSC) 60, 63–4
  fast component 66
  slow component 65–6

felodipine 38
field-effect transistors (FET) 8, 9
fluctuation analysis
  glutamate-activated conductances 56–8
  stretch-activated ion channels 76–7
fluoride ions (F$^-$) 33, 139
fura-2 66–7

GABA$_A$ receptors 183–202
  antibodies, *see* anti-GABA$_A$ receptor antibodies
  expression in *Xenopus* oocytes 162, 183–4
  in situ hybridization 206, 222
  subunit structure 183–5
gadolinium (Gd) 39
GDEE (glutamic acid diethyl ester) 53
GDP-β-S (guanosine 5'-O-(2 thio)-diphosphate) 101, 108
gigaohm seals (giga-seals) 6–7, 82, 177
glasses, patch-pipette 21–2
glial cells
  glycine release 52–3
  primary cultures, agonist-induced InsP formation 122
glutamate 33, 103, 104
L-glutamate 52, 53, 58, 59, 60, 63
glutamate receptors 49–69
  concentration clamp techniques 61–4
  experimental preparations 50–1
  future directions 69
  kinetics of synaptic responses 64–6
  metabotropic 53, 67–8, 69
  oocyte expression 51, 68–9, 162
  optical measurements 66–8
  pharmacology 52–3
  single-channel recordings 58–61
    cell-attached configuration 61
    concentration clamp techniques 63–4, 65
    kinetic analysis and multichannel patches 60
    NMDA channels 58–9
    non-NMDA channels 60
  whole-cell recordings 53–8, 61–3, 65
    concentration clamp techniques 61–3, 65
    fluctuation analysis technique 56–8

  NMDA channel blockers 56
  voltage-dependent conductances 55–6
  see also kainate receptors; *N*-methyl-D-aspartate (NMDA) channels/receptors; non-NMDA channels/receptors; quisqualate receptors
glutaraldehyde method for coupling of peptide and carrier protein 189, 191
glycerophosphate 139
glycine, activation of NMDA receptors 49, 52–3, 61–2
glycine-gated channels, expression in *Xenopus* oocytes 162
G protein-coupled receptors
  binding studies 100
  criteria for identification 99
G protein-mediated processes
  criteria for identification 97–9
  experimental strategies for identification 99–109
G proteins 95–110
  activated exogenous 106–7
  antibodies 105–6, 108
  bacterial toxin sensitivity 96, 102–5
  characteristics of systems linked to 96–7
  classification 95–6
  cloned 107
  G$_i$ 95, 100, 101, 102, 103, 106
  G$_k$ 106
  glutamate receptor-linked 69
  G$_o$ 95, 97, 102, 103, 106
  G$_s$ 95, 96, 97, 101, 106
    mutants 107
    receptor coupling 100
    toxin sensitivity 102, 103
  GTPase activity 100–1
  G$_T$ (transducin) 95, 102
  guanine nucleotide analogues, effects of 101–2
  mutant 107
  photoaffinity labelling 102
  reconstitution studies using exogenous 106–7
  subunits 95–7
    reconstitution studies 106–7
growth cones, stretch-activated ion channels 76
GTP 34, 95, 96, 97, 106
  G protein-coupled receptor studies 100
  intracellular photo-release 108–9
GTPase activity of G protein 100–1
GTP-binding proteins, *see* G proteins
GTP-γ-S (guanosine 5'-O-(3 thio)-triphosphate) 101, 106, 108
  intracellular photo-release 108–9, 110
guanine nucleotide analogues
  intracellular photo-release 108–9, 110
  modification of G protein activity 101–2

231

Index

guanine nucleotide analogues (*cont.*):
  permeabilization of cells 107–8
  photoaffinity-labelled 102
  whole-cell recording studies 108
guanylylimido-diphosphate (GMP–PNP, GppNHp) 100, 101, 108
guanylyl (β,γmethylene)-diphosphate 101

hair cells 76
Hepes 33, 41, 118
high performance liquid chromatography, *see* HPLC
H-infinity plot, Ca^{2+} channels 37, 39
hippocampus, long-term potentiation 49, 68
holders, patch-pipette 23–4
holding potential (HP) 34–5, 36
HPLC (high performance liquid chromatography)
  inositol phosphates 129–30, 131
  synthetic peptide calcineurin substrate 147–8
hybridization, *in situ*, *see in situ* hybridization
hybridization buffer, oligonucleotide 214, 221, 223, 224
hydrochloric acid (HCl) 125
hydrostatic pressure
  activation of stretch-activated ion channels 77–8
  sensitivity of stretch-activated ion channels 78–81
hypo-osmotic shock
  activation of stretch-activated ion channels 85–6
  permeabilization of cells 108

immunoblots, anti-$GABA_A$ receptor antibodies 195–7
immunoprecipitation assay, anti-$GABA_A$ receptor antibodies 199–202
inactivation, Ca^{2+} channels 29, 36, 39
inactivation rates, *see* decay-times
inhibitor-1 and inhibitor-2 152
inositol 115–16
  radiolabelled ([^{3}H]inositol) 116
  labelling tissue slices with 119–22
  sources and purification 119
inositol-1,4-biphosphate (Ins[1,4]P_2) 115, 130, 131
inositol cyclic phosphates 125, 126
inositol-1-monophosphate (Ins[1]P) 115, 116, 130, 131
inositol phosphates (InsP) 115–16
  measurement of agonist-induced formation 115–31
    buffer composition 118
    extraction methods 125–6
    incubation with agonist 123–5

    incubation with LiCl and antagonists 122–3
    labelling slices with [^{3}H]inositol 119–22
    measurement of Ins[1,4,5]P_3 and DAG mass 130–1
    preparation of slices 118–19
    protocol 116–17
    separation on Dowex columns 126–8
    separation using HPLC 129–30
  non-receptor-mediated increases in formation 124–5
inositol-1,3,4,5-tetrakisphosphate (Ins[1,3,4,5]P_4) 116, 126, 130, 131
inositol-1,3,4-trisphosphate (Ins[1,3,4]P_3) 116, 130, 131
inositol-1,4,5-trisphosphate (Ins[1,4,5]P_3) 115, 123, 124
  measurement of mass 130–1
  metabotropic glutamate receptors and 67–8, 69
  separation on Dowex columns 130, 131
in situ hybridization 205–23
  autoradiography of sections 219
  controls 219–23
  future prospects 223
  hybridization of probes to tissue 217–19
  materials and reagents 213–14
  preparation and fixation of tissue 215–17
  scheme of procedure 206, 207
  sterilization of solutions 211–12
  synthetic oligonucleotide probes 205–11, 223
  types of probes used 205
insulin 151–2
integrating headstage technique, *see* capacitor feedback technique
ion channels
  glutamate-gated 52
    see also N-methyl-D-aspartate (NMDA) channels/receptors; non-NMDA channels/receptors
  G protein-linked 97–9
  RNAs
    injection with oocytes 174–5
    synthesis and isolation 162–71
  stretch-activated, *see* stretch-activated ion channels
  Xenopus oocytes
    electrophysiology 175–80
    endogenous activity 178–9
    exogenous interfering 179–80
    expression of exogenous 161–2
  see also calcium channels; chloride channels; potassium channels; sodium channels
ion exchange chromatography (Dowex columns)
  calcineurin assay 148–9

Index

preparation 126–7
separation of inositol phosphates 126–8
ion selectivity
  Ca^{2+} channels 30
  NMDA channels 58–9
  stretch-activated ion channels 84, 86–9

kainate receptors 52, 54–5, 67
  expression in *Xenopus* oocytes 69, 162
  single-channel recordings 60
  voltage sensitivity 55
ketamine 56
keyhole limpet haemocyanin (KLH) 188–91
kinases, *see* protein kinases
kinasing, synthetic oligonucleotides 209
Kreb's bicarbonate buffer (KRB) 118

lens, frog, stretch-activated ion channels 81
leupeptin 34, 46
lithium chloride (LiCl)
  agonist-induced formation of InsP and 116–17, 122–3
  RNA isolation 164, 165
long-term potentiation (LTP) 49, 68

magnesium ions (Mg^{2+})
  blockade of NMDA channels 49, 52–3, 55–6, 59, 67
  G proteins and 97, 101
  radiolabelling of phosphoproteins 136–7, 138
  sensitivity of Ca^{2+} channels 39
m-maleimidobenzoic acid N-hydroxysuccinimide ester (MBS)
  method for coupling of peptide and carrier protein 190–1
MAP 2 protein 205
mechanoreceptors 76
mechano-sensory cells 76
metabotropic glutamate receptors 53, 67–8, 69
MK801 56, 57, 59
mole fraction effect 36
mRNA, *see* RNA
muscarinic acetylcholine receptors 162, 179
muscle
  chick, stretch-activated ion channels 81, 89
  crayfish 63
myosin light-chain kinase 141

NBQX (2,3-dihydroxy-6-nitro-7-sulphamoyl-benzo(F)quinoxaline) 53
N-ethylmaleimide 105
neurons
  acutely dissociated, glutamate receptors 50, 51, 61
  dissociated cultures, glutamate receptors 61, 65–6

growth cones, stretch-activated ion channels 76
primary cultures
  agonist-induced InsP formation 122, 130, 131
  glutamate receptors 50–1, 58–9
neurotensin receptors 162
neurotransmitter receptors
  expression in *Xenopus* oocytes 161–2
  RNAs, synthesis and isolation 162–71
nickel ions (Ni^{2+}) 29, 31, 39
nicotinic acetylcholine receptors 60, 162
nifedipine 30, 46
nisoldipine 30
nitr-5 68
nitrendipine 30
nitrosoacetophenone 109
N-methyl-D-aspartate (NMDA) channel blockers
  single channel analysis 59
  whole-cell analysis 56, 57
N-methyl-D-aspartate (NMDA) channels/receptors 49
  concentration clamp technique 61–4
  expression in *Xenopus* oocytes 69, 162
  in vitro preparations 51
  noise analysis 58
  optical methods 67–8
  pharmacology 52–3
  rundown 54
  single-channel recordings 58–9, 60, 63–4
  synaptic activation 65–6
  voltage-dependent conductances 55–6
  whole-cell recordings 54–8, 61–3
noise
  patch-pipette glasses 21
  single-channel patch-clamp technique 6–7, 11–13
noise analysis, *see* fluctuation analysis
non-NMDA channels/receptors
  concentration clamp technique 63–4
  pharmacology 52, 53
  single-channel recordings 60, 63–4
  synaptic activation 65–6
  see also kainate receptors; quisqualate receptors
norepinephrine (noradrenaline), InsP formation 120, 123, 130, 131
Np-methyl-D-glucamine (NmDg) 33

octanol 31
okadaic acid 138, 152
oligo(deoxyribo)nucleotide probes, synthetic 205–6
  artefacts in tissue binding 219–23
  autoradiography 219
  hybridization to tissue 217–19
  non-radioactively labelled 223

Index

oligo(deoxyribo)nucleotide probes (cont.):
  purification 208–9
  radioactive labelling 209–11
    kinasing method 209
    terminal transferase reaction method 209–11
    radiolabelled, analysis on acrylamide gels 211, 212
  synthesis 206–8
oligo(dT) cellulose 164–5
ω-conotoxin (ωCgTx) 46
  sensitivity of Ca^{2+} channels 29, 31, 38, 45
oocytes, *Xenopus*, see *Xenopus* oocytes
opossum kidney (OK)-cells 81, 85–6, 87, 88
optical methods, glutamate receptors 66–8
orthovanadate 139

papain 51
patch-clamp techniques
  concentration clamp technique 63–4, 65
  G protein-coupled ion channels 97–9, 108
  instrumentation technology 3–24
    capacitor feedback technique 9–15
    measuring changes in membrane capacitance 15–16
    micropipettes, *see* patch-pipettes
    resistor feedback technique 7–9, 14–15
    seal formation 6–7
    series resistance compensation 16–20
  loose 4
  Xenopus oocytes 176, 177–8
  see also single-channel recordings; whole-cell recordings
patch-pipettes 4, 6, 21–4, 177, 178
  big patches 177–8
  glasses 21–2
  holders 23–4
  pullers 22–3
peptides, synthetic
  coupling to carrier proteins 188–91
  for production of anti-GABA$_A$ receptor antibodies 185–91
  protein phosphatase substrate 146, 147–9
  HPLC purification 148–7
  phosphorylation 148
permeabilization of cells 107–8
pertussis toxin 96, 102, 103–5
pH buffers, for studying Ca^{2+} channels 33, 34, 41
phencyclidine 56
phenothiazines 152
phorbol esters 152
phosphatases, protein, *see* protein phosphatases
phosphate 33
phosphatidic acid (PA) 115
phosphatidylinositol (PI) 115, 120–1

phosphatidylinositol-4-phosphate (PIP) 115, 120–1
phosphatidylinositol-4,4-bisphosphate (PIP$_2$) 115, 120–1
phosphoinositide hydrolysis
  measurement of 115–31
  non-receptor-mediated increases 124–5
phospholipase A$_2$ 96, 106
phospholipase C, phosphoinositide-specific 115
phosphoproteins 135
  characterization of phosphorylation sites 139
  detection 136–9
  establishing a physiological role 151–4
  phosphate analyses 139
  primary structure 143
  radiolabelling with ^{32}P 136–9
  as substrates for protein kinases 142–3
  as substrates for protein phosphatases 146, 149
phosphorus-32 (^{32}P)
  labelling of phosphoproteins 136–9, 146–7, 148
  labelling of synthetic oligonucleotides 209–11
phosphorylation of proteins, *see* protein phosphorylation
photo-affinity-labelled guanine nucleotide analogues 102
photo-release of guanine nucleotide analogue 108–9, 110
phytic acid hydrolysate 126
piezoelectric translator 63, 64, 65
placental protein tyrosine phosphatase 1B (PTPase1B) 153
plasmid cloning vectors 166
poly-L-lysine coating of slides 216
poly(U) Sepharose 165
positive feedback (correction) technique of series resistance compensation 16–18
potassium (K$^+$) channels
  blockers 33, 41
  endogenous, *Xenopus* oocytes 179
  expression in *Xenopus* oocytes 162
  G protein-coupled 106
potassium ions (K$^+$)
  role in volume regulation 85, 86–8
  stretch-activated ion channels selective for 76
prediction (supercharging) technique of series resistance compensation 18–20
preprosomatostatin 224
protease 82
protein kinase 141
  multifunctional calmodulin-stimulated, *see* calmodulin kinase II
protein kinase C 89, 115

Index

ein kinases 135–6
says 140–3
tablishing a physiological role 151–4
entification 139–45
hibitors 136, 139, 141
   mammalian neural tissues 141
imary structure of substrate 143
rine/threonine 139
rosine 139
e also calmodulin kinase II
ein phosphatase 1 (PP1) 152, 153
ein phosphatase 2A (PP2A) 153
ein phosphatase 2B (PP2B) 153
ein phosphatase 2C (PP2C) 153
ein phosphatases 135–6
says 146–9
tablishing a physiological role 151–4
entification 145–50
hibitors 136, 138, 139, 152, 153
   mammalian tissues 153
rine/threonine-specific 145, 153
rosine-specific 145–6, 153
e also calcineurin
ein phosphorylation 135–6
ochemical approach to study 136
tablishing a physiological role 151–4
sting for effects on function 150–1
e also phosphoproteins; protein kinases;
   protein phosphatases
ein tyrosine phosphatase 1B (PTPase1B),
   placental 153
ein tyrosine phosphatase 5 (PTPase 5),
   brain 153
rs, patch-pipette 22–3
-chase labelling with [^{3}H]inositol 122
e protocols, Ca^{2+} channels 34–6, 42
phosphate 139

tz micropipettes 21–2, 23
oxalinediones 53
qualate receptors 52, 63, 162
gle-channel recordings 60
naptic activation 66
ltage sensitivity 55

ene products 95, 96, 102, 107
tor feedback patch-clamp headstage 7–9
opsin kinase 141
nuclease A (RNAse A) 221, 222
nucleases, preventing degradation of
   RNA by 163–4, 211–12
(mRNA)
pression in *Xenopus* oocytes 68–9,
   161–2
jection into oocytes 174–5
situ hybridization 205–23
lation and purification 164–6
eventing degradation 164–5, 211–12

size fractionation 167–71, 172
synthesis *in vitro* 162, 166–7
rundown
   Ca^{2+} channels 34, 36
   NMDA receptors 54

saline, physiological 42
salmon sperm DNA, preparation 213–14
sapphire 22
seal formation, patch-clamping 6–7, 82, 177
SEP-PAK purification of synthetic
   oligonucleotides 208–9
series resistance compensation 16–20
   positive feedback 16–18
   supercharging (prediction) 18–20
serotonin (5-HT), InsP formation 120, 121,
   123, 124
serotonin (5-HT) receptors 162
single-channel recordings 3, 4, 5–6
   Ca^{2+} channels 29, 32, 40–5
   concentration clamp technique 63–4, 65
   glutamate receptor-linked channels 58–61,
      63–4, 65
   instrumentation technology
      capacitor feedback technique 9–15
      micropipetts, see patch-pipettes
      resistor feedback technology 7–9
      seal formation 6–7
   ion channels expressed in *Xenopus* oocytes
      176, 178
   stretch-activated ion channels 76–7, 82
site-directed mutagenesis, G proteins 107
slides, poly-L-lysine coating 216
sodium (Na^+) channels
   eel 161, 162
   endogenous, *Xenopus* oocytes 179
   expression in *Xenopus* oocytes 162
   inhibition 33, 41
sodium ions (Na^+), substitution 33
somatostatin receptors 162
soybean trypsin inhibitor 188
spinal cord neurons, glutamate receptors
   58, 63–4
streptolysin O 108
stretch-activated (SA) ion channels 75–89
   activation during volume regulation 84–9
   crayfish stretch-receptors 76, 81–4, 85, 89
   hydrostatic pressure and membrane tension
      77–8
   physiological function 75–6
   pressure and voltage-dependence of channel
      open probability 78–80
   pressure- and voltage-sensitivity in different
      cell types 80–1
   recording techniques and fluctuation
      analysis 76–7
   Xenopus oocytes 178
stretch receptors, crayfish 76, 81–4, 85, 89

Index

strontium ions (Sr^{2+}) 31, 32, 41
substance K receptors 162
substance P receptors 162
sucrose density gradients, RNA size fractionation 168–70
suction
  activation of stretch-activated ion channels 78, 82–4, 85–6
  patch-clamping of *Xenopus* oocytes 177
sulphur-35 (^{35}S), labelling of synthetic oligonucleotides 209–11
supercharging technique of series resistance compensation 18–20
synapsin-1 144
synpatic responses, glutamate receptor activation 64–6

tails (deactivation), Ca^{2+} channels 29–30, 36
terminal transferase reaction 209–11
test potentials (TPs) 34–6
tetraethylammonium (TEA) 33
tetramethrin 31
tetrodotoxin (TTX) 33, 41, 42
tissue slices
  agonist-induced InsP formation 118–19, 120–1, 123, 124
  glutamate receptors 50, 51
toxins, bacterial 96, 102–5
transducin (G_T) 95, 102
trichloroacetic acid (TCA) extraction, inositol phosphates 125–6

urea 164, 165

vasopressin receptors 162
verapamil 31
voltage-clamp technique 3–4, 54
  ion channels expressed in *Xenopus* oocytes 175–7

voltage sensitivity
  glutamate-gated channels 55–6
  stretch-activated ion channels 80–1
volume regulation 75–6
  role of cytoskeleton 89
  stretch-activated ion channel activation 84–9

whole-cell recordings 3, 4–5
  Ca^{2+} channels 31–40
  glutamate-gated channels 53–8, 61–3,
  G protein-coupled ion channels 108
  instrumentation technology
    capacitor feedback technology 13–1
    measuring changes in membrane capacitance 15–16
    patch-pipettes, *see* patch-pipettes
    resistor feedback technology 7–9
    series resistance compensation 16–2
  perforated patch method 55
  voltage-clamp technique 4, 54

Xenopus oocytes 161–80
  electrophysiology of ion channels 175-
    big patches 177–8
    isolation of exogenous currents 178-
    patch clamping 176, 177–8
    single-channel recording 176, 178
    two-electrode voltage clamping 175-
  endogenous ion channel activity 178–
  expression of neurotransmitter recepto and ion channels 161–2
  $GABA_A$ receptor expression 162, 183
  glutamate receptor expression 51, 68–
  injection and culture 174–5
  isolation and preparation 172–4
  RNA synthesis and purification 162–7

zinc ions
  effect on NMDA responses 59